Water Supply Systems and Evaluation Methods

Volume II: Water Supply Evaluation Methods

Harry E. Hickey, Ph.D.

Acknowledgement

This project was carried out by the Society of Fire Protection Engineers (SFPE) and was supported by the Department of Homeland Security's (DHS's) Science and Technology Directorate and the United States Fire Administration (USFA). SFPE is an engineering association for advancing the science and practice of fire protection engineering. Water supply is an important subject to the fire service, fire protection engineers, and city managers. These manuals are intended to provide a reference for concepts and terminology to facilitate communication and understanding among these organizations.

About the Author

Dr. Harry E. Hickey's career in fire protection spans more than 50 years. He taught Fire Protection Engineering at the University of Maryland for 26 years. He also has extensive experience in the fire service as a firefighter, fire officer, and emergency coordinator. His combination of municipal fire administration and fire protection engineering experience gives him unique insight into the challenges of design and operation of municipal water supplies.

He received his Ph.D in Public Administration from American University in Washington, DC. He has authored many book and articles including *Public Fire Safety, A Systems Approach*; *Fire Protection Hydraulics*; and two editions of *The Fire Suppression Rating Schedule Handbook*.

Table of Contents

Chapter 1: Evaluating Municipal Water Distribution Systems

Primary Considerations

This topic has two primary objectives: The first is to understand the evaluation of an installed municipal water supply delivery system by identifying all the physical components of any specific water distribution system. The same basic concepts and principles apply to small community water systems and large city water systems. For a basic understanding of these concepts, two illustrations are provided that include a relatively small water distribution system and a medium size water distribution system. These concepts will cover 92 percent of all the water supply systems in the United States. While there are similarities to all water systems, it should be recognized that the likelihood of two water systems being exactly alike in physical features is remote because the raw water sources in relation to the water delivery demands can hardly be the same.

The second objective is to provide recognized practices for conducting water supply tests at prescribed intervals to measure the water system delivery capability and ensure that the system is meeting the water supply demand. An important part of this second objective is to use the results of water supply tests to monitor the performance of the water delivery system in relation to the existing water supply and the constant changes in demand on the water system. The following material will illustrate the broad features of water supply systems in order to understand how this can be accomplished. Chapter 2 presents a basic understanding of hydraulic fundamentals needed to accomplish water supply testing and evaluation accurately, and Chapter 3 presents water supply system evaluation methods for determining existing water supplies for consumer consumption and especially for fire protection

Functional Components of a Water Utility System

A water utility system can be relatively simple for a community of 3,000 to 5,000. At this level of population, communities often are served by wells. The well water is treated typically by chlorination and then either pumped directly into water distribution mains to supply customers or pumped into ground-level or elevated storage tanks where the water flows by gravity on demand to each customer on the water system. Some fire hydrants **may be** located on the distribution system to provide a minimum fire flow capability in the range of 250 gallons per minute (gpm) to 500 gpm. **Figure 1-1** illustrates an actual example of a community that has these characteristics.

Figure 1-1

Features of a Small Community Water Distribution System

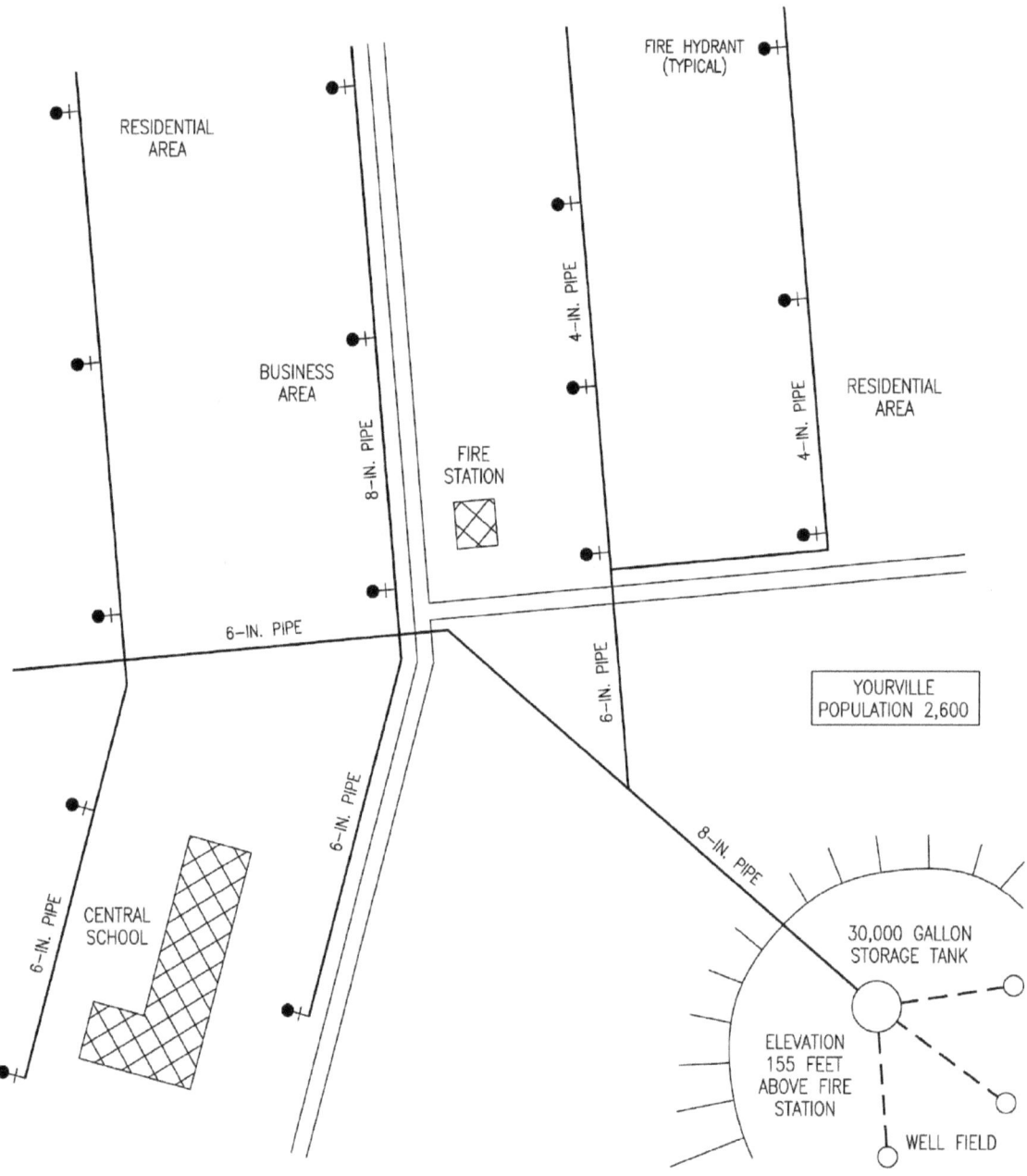

As the population served in the illustration of a small community increases, so does the complexity of the water delivery system. **Figure 1-2** depicts the functional components expected to be in place in communities with populations ranging from 25,000 to 50,000. (Reference #1, pg. 12) This is fairly typical of what one needs to understand in evaluating water supply systems to assure that rates of water can be delivered through the distribution system to simultaneously meet consumer consumption demands and meet Needed Fire Flow (NFF) criteria for structural fire suppression. Therefore, before specifically examining and tracing the water system component diagramed in **Figure 1-2**, it is essential that each component be evaluated in relationship to the capability of the water delivery system, plus the function of selected components of the water system to meet the needed domestic and fire protection demands on the system. This needs to be assessed for each and every water system as a function of **rates of water usage**. Three historical or predicted water demand rates are involved in the discussion of consumer demand and fire protection: (Reference #2, pg. 12)

- **Average daily demand**–the average of the total amount of water used each day during a 1-year (designed) period.

- **Maximum daily demand**–the maximum total amount of water used during any 24-hour period. The Insurance Services Office. Inc. (ISO) bases this calculation on the highest demand during the previous 3 years from the years of an ISO *Grading Schedule* evaluation. Note: This number should consider and exclude any unusual and excessive uses of water that would affect the calculation i.e., a broken water main.

- **Maximum hourly demand**–the maximum amount of water used in any single hour of any day in a 3-year period. It normally is expressed in gallons per day by multiplying the actual peak hour by 24.

Figure 1-2

Features of a Minimum Size Community Water Distribution System

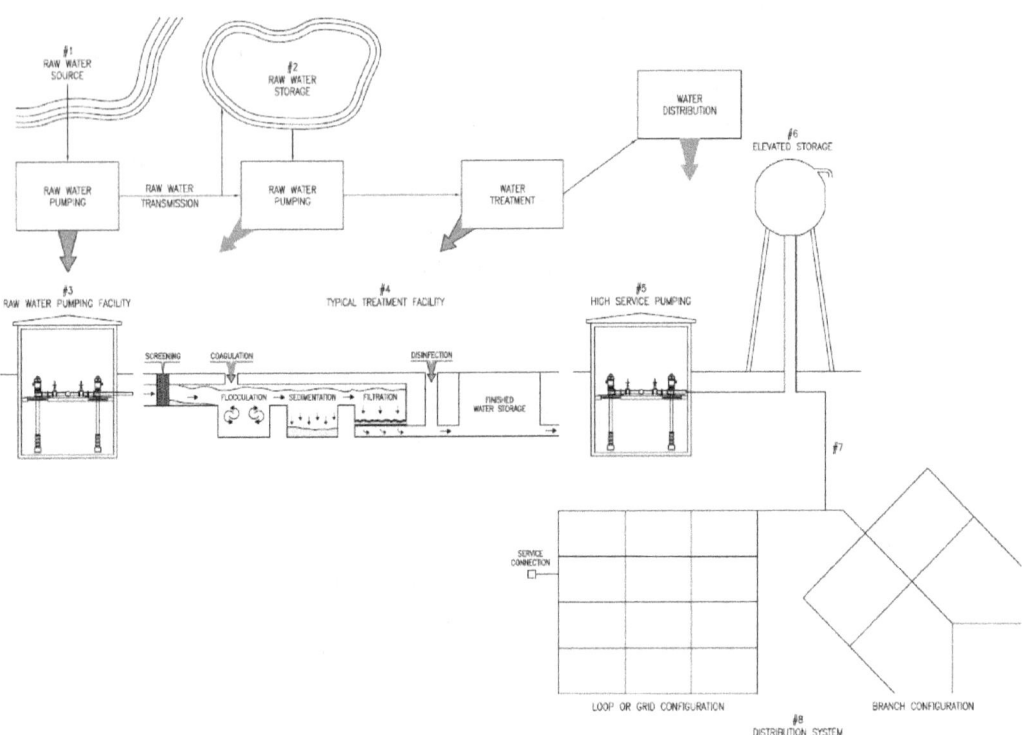

When specific data on past consumption levels are not available, a good rule of thumb is that maximum daily demand may vary from 1.5 to 3.0 times the average daily demand, while the peak hourly rate may vary from 2 to 8 times the average daily rate. In small water systems, peaking factors may vary significantly higher.

Design flow and analysis should be based on the maximum hourly demand or the maximum daily demand plus the fire flow requirement, whichever is greater. This distribution system should be designed to maintain a minimum pressure of 20 pounds per square inch (psi) at **all** water taps including fire hydrant locations under all conditions of design flow.

[This is a recommended practice of the American Water Works Association (AWWA) *Manual of Water Supplies Practices*–M-31 and M-17 along with criteria published by the ISO in accordance with Grading Schedule evaluations.]

In order to account for system demands, chart recorders should be in place at every separate location where **purified water** enters the distribution system, including finished water holding basins, direct pumping facilities, finished water standpipe tanks, and finished water gravity tanks. This is the only reasonably accurate way to monitor water system demand on an hourly, or less frequent, time period, 24 hours a day, 365 days a year. This is a key consideration in matching water system demand to water system availability.

TRACING THE COMPONENTS OF AN URBAN WATER DISTRIBUTION SYSTEM

The following information is presented in the context of **Figure 1-2.** (Reference #1, pg. 12)

1. Raw water source shown in the upper left of the drawing marked #1.

   The water source may be a lake, a river, a reservoir, a well field, or, more recently, salt water sources which can be purified through new techniques in water filtration and reverse osmosis. The Tampa, Florida, municipal water system now has the capability of producing 60 percent of the city's water demand using salt water. The AWWA recommended guideline on raw water capability is that the supply source(s) have a sufficient capacity at all time to meet maximum daily demand for a continuous period of 5 days. (Reference #1, pg. 12) This is demonstrated on the drawing by a raw water storage pond marked #2.

   Water Department action needed: A depth gauge is needed on each raw water source. A hydrologist or geologist needs to asses the gallons of water in storage at gauge level. There typically is not an equal drop or increase as water is drawn off or is supplemented by rain and runoff, due to the slope of the container sides and the slope of the bottom of the holding basin. This monitoring needs to be done daily if a diminishing supply is observed, to predict long-term supply conditions.

2. Typical raw water pumping facility marked #3.

   In the case of the illustrated water system, the pumping facility has a dual purpose. First it can pump the raw water, which is filtered by trash racks (coarse screws) and other finer screening, if necessary, to the treatment facility where the water is processed to meet Environment Protection Agency (EPA) criteria and even more rigid requirements for water treatment in several States through State health departments. Second, there needs to be the capability to transfer water to and from the raw water storage facility. This builds reliability into the water system. Constant-recording flow meters are needed on each of the pumps in this facility to assess how and when water is being transported through facility and the rate of water in gpm.

   Water Department action needed: The secondary raw water storage facility is important for retaining a reserve supply of water in case of a major pipe failure on the distribution piping, or if the main source of water become depleted due to drought conditions or contamination of the primary supply.

3. The treatment facility marked #4.

   Chapter 6 covers the basic of water supply treatment and the sampling of water required to meet EPA criteria. There is a need to know the maximum water processing rate and the length of time that water can be processed at this rate, because this could limit water delivery to the distribution piping system. Process flow rates need to be monitored on a continuous timeline, which also will account for the downtime in the treatment plant to flush and clean equipment.

   Note in the illustration that a **finished water** facility is provided on the delivery side of the treatment plant; this is commonly called a clear well. In gravity feed systems, water flows from the clear well(s) into the distribution piping, or it is pumped where the land surface is relatively level. Water levels in the clear well(s) needed to be monitored closely on a daily basis with data recorded hourly.

4. A high service pumping station marked #5.

    Note that this pumping station is located on the water distribution delivery side of the water treatment plant. High-level service pumps may be needed to:

    a. Pump water up to service areas that have higher elevations than other areas of a community.

    b. Fill gravity tanks that float on the water supply distribution system.

    c. When service pumping stations are used to distribute water, and no water storage is provided, the pumps force water directly into the water mains. From a water system evaluation perspective, there is no outlet for the water except as it furnishes consumer consumption for actual fire flows. Variable speed pumps or multiple pumps may be required to provide adequate water delivery service because of fluctuating demands. The efficiency and expense of this pumping equipment needs to be considered carefully. For example, it is a disadvantage that the peak power demand of the water plant is likely to occur during periods of high electrical consumption, and thus increase power costs. Furthermore, systems with little or no storage should be provided with standby electrical generating capability or pumps driven directly by internal combustion engines. These standby generators and engines needs to be tested routinely (e.g., several hours per week).

5. A gravity storage facility marked #6.

    An extremely important element in a water distribution system is water storage. (Reference #3, pg. 12) System storage facilities have a far-reaching effect on a system's ability to provide adequate consumer consumption during periods of high demand while meeting fire protection requirements. The two common storage methods are ground-level storage and elevated storage. The finished water storage at number 4 on **Figure 1-2** is an example of ground-level storage; this type of storage also may be contained in covered tanks. Emphasis is put on elevated storage as a stand-alone in Chapter 7 of this Manual.

6. Water entering the distribution system marked #7.

    There are two basic types of pipe layout for delivering water to consumer taps and to supply water to individual fire hydrants. The preferred method is to loop the entire service area with a primary feeder main; the size is determined by hydraulic analysis. Interior to the ring main are cross-connected secondary feeders provided along the major streets in the community. Interior water mains that essentially provide water to residential areas are cross-connected to the secondary feeders. The advantages of this type of pipe system layout are two fold: 1) the water to every service location or demand point is supplied from two directions, which is considered to be the most efficient hydraulic design to minimize pipe sizes; and 2) in the event that a pipe section is out of service for cleaning, breakage from an accident, tapping for service extension, or whatever reason, water can be supplied to any demand point from a different travel path. In older community water systems a single primary feeder supplies secondary feeders and distributor pipes along block fronts in a branched layout configuration where at any demand point water is supplied from one direction only. This arrangement decreases the reliability of a water system significantly and has a tendency to decrease fire flow capability for larger scale fires. The capability of water main systems for meeting fire flow criteria can be determined only by semiannual fire flow tests as presented below.

SPECIFIC CONSIDERATIONS

Figure 1-2, discussed above, provides a conceptual layout of a typical water distribution system and highlights the features of this system that require evaluation on a frequency ranging from hourly to annually. A number of water system features need to be examined more specifically in relation to providing adequate and reliable water supplies for fire protection. The following topics need to be considered carefully in the evaluation of any given water supply system, not only for consumer consumption but especially for meeting model building code required water supplies and/or the ISO's NFF for public-sector fire protection.

Distribution System Storage: An extremely important element in a water distribution system is water storage. System storage facilities have a far-reaching effect on a water system's ability to provide adequate and reliable water supplies for domestic needs and especially for fire protection. Storage within a distribution system enables the system to process water at times when treatment facilities otherwise would be idle. It then is possible to distribute and store water at one or more locations in the service area which are closer to the end user, and provide the needed volume of water at a minimum of 20 psi residual pressure on fire hydrants to meet fire flow demands. The considerations presented below need to be evaluated carefully for any given municipal water delivery system.

1. Advantages: The principal advantages of distribution storage include the fact that storage equalizes demands on supply sources, production works, plus transmission and distribution mains. As a result, the sizes or capacities of these elements need not be so large. Additional system flows and pressures are improved and stabilized to better serve the customers throughout the service area. Finally, reserve supplies are provided in the distribution system for emergencies, such as fire suppression and power outages.

2. Meeting system demands and required/NFF: In the evaluation of both existing and needed water supply storage, it is essential to consider the location, capacity, and elevation (if elevated) of distribution storage in relation to system demands and the variation of demands that occur throughout the day in different parts of the system. System demands can be determined only after a careful analysis of an entire distribution system. However, some general rules may serve as a guide to such an analysis. **Table 4-1** in Chapter 4 lists daily and hourly variations for a typical city and the resultant storage depletion. (Reference #4, pg. 12) Such data are of great assistance in determining the adequacy of storage capacities. This type of analysis needs to be performed for each water system because each has its own requirements. Studies show that it is more advantageous to provide several smaller storage units in different parts of the water system than to provide an equivalent capacity at a central location. Smaller pipelines are required to serve decentralized storage and, other things being equal, a lower flow-line elevation and pumping head or pressure results.

3. Comparing and contrasting ground storage and elevated storage: Storage within the distribution system network normally is provided in one of two ways: ground-level storage with high-service pumping, and elevated storage. It is noted by the AWWA that elevated storage provides the best, most reliable, and most useful form of storage, **particularly** for fire protection. (Reference #5, pg. 12)

a. Ground-level storage: Since water kept in ground-level storage is not under pressure, it must be delivered to the point of use by pumping equipment. This arrangement limits water system effectiveness for fire protection in three ways: 1) There must be sufficient excess pumping capacity to deliver the peak demand for normal uses as well as any fire demands; 2) Standby power sources and standby pumping units must be maintained at all times because the system cannot function without the pumps; and 3) The distribution lines at all points in the system must be significantly oversized to handle peak delivery use plus fire flow, no matter where a fire might occur.

b. Elevated storage: Properly sized elevated water tanks provide dedicated fire storage and are used to maintain constant system pressure. Where elevated tanks are used, ground storage tanks **may** still be required to provide water to the water treatment plant based on the type of system design. However, the size of the ground storage tank(s) can be reduced to the minimum required to treat the water adequately, usually in the 300,000- to 500,000-gallon range. The elevated storage tanks are used to store treated water to provide water directly to the water distribution system. What needs to be evaluated with elevated storage is that the water supply is fed to the distribution system from the top 10 to 15 feet of water in the elevated tank(s). The high-service pumps are constant-speed pump units, which can operate at their highest efficiency point virtually all the time. The remaining water in the tanks (70 to 75 percent) normally is held in reserve as dedicated fire storage. When domestic consumption is at a minimum during a 24-hour period, water in the gravity tank is recycled with fresh treated water to eliminate aging of the water in the holding tank. (Reference #6, pg. 12)

4. Pumping for distribution storage: The two types of distribution storage (i.e., ground level and elevated) have, in turn, two types of pumping systems that need to be evaluated on a performance basis. One is a direct pumping system, in which the instantaneous system demand is met by pumping with no elevation storage provided. The second type is an indirect system in which the pumping station lifts water to a reservoir or elevated storage tank, which floats on the water system and provides system pressure by gravity. Some comments on each pumping system are in order:

a. Direct pumping: The direct pumping system is being phased out for municipal water delivery systems primarily because of operating costs. However, many older systems, especially for smaller communities, still exist. Variable-speed pumping units operated off direct system pressure are also in use in some communities. Hydropneumatic tanks at the pumping station provide some storage. These tanks permit the pumping-station pumps to start and stop, based on a variable-system pressure preset by controls operating off of the tank.

b. Indirect pumping: In an indirect system, the pumping station is not associated with the demands of the major load center. It is operated from the water level difference in the reservoir or elevated storage tank, enabling the prescribed water level in the tank to be maintained. The majority of systems have an elevated storage tank or reservoir on high ground floating on the system. This arrangement permits the pumping station to operate at a uniform rate, with the storage either making up or absorbing the difference between station discharge and system demand.

EVALUATING DISTRIBUTION SYSTEM APPURTENANCES

1. Piping and valve arrangement.

   A piping system serving the consumers in a small community is illustrated in **Figure 1-3**. The primary feeders, sometimes called arterial mains, form the skeleton of the distribution system. They are located so that large quantities of water can be carried from the pumping plant to and from the storage tanks and the distribution system. (Reference #7, pg. 12)

   Primary feeders should be arranged in several interlocking loops, with the mains not more than 3,000 feet apart. Looping allows continuous service as previously identified through the rest of the primary mains, even when one portion is shut down temporarily for repairs. Under normal conditions, looping also allows supply from two directions for large fire flows. Large feeders and long feeders should be equipped with **blow-off valves** at low points and **air relief valves** at high points. Valves should be placed so that a pipe break will affect water service **only** in the immediat area of the break.

Figure 1-3

Typical Small City Distribution System

(Reference #7, pg. 12)

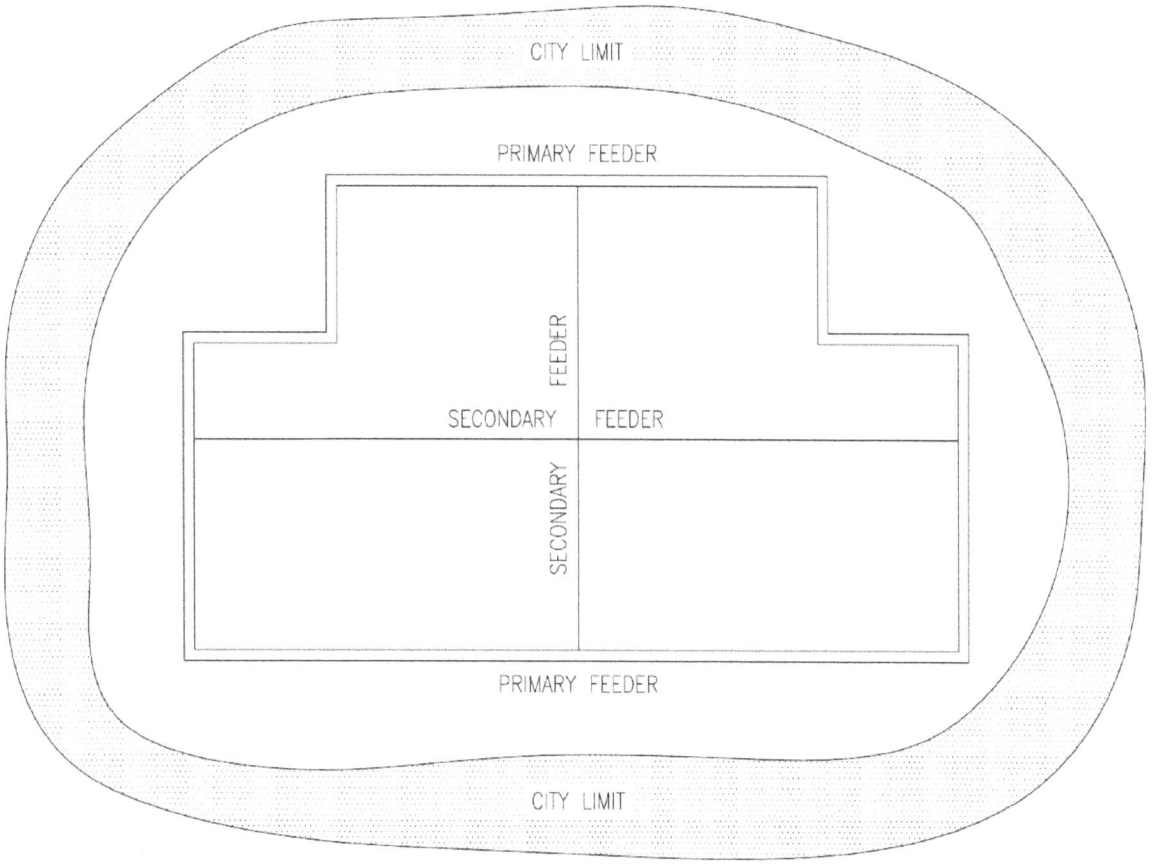

The secondary feeders carry large quantities of water from the primary feeders to points in the system in order to provide for normal domestic consumption supply and fire suppression. They form smaller loops within the loops of the primary mains by running from one primary feeder to another. Secondary feeders should be spaced only about three blocks apart, or a maximum of 1,500 feet. This spacing allows concentration of large amounts of water for firefighting without excessive head loss and resulting low pressure.

Small distribution mains (i.e., distributors) are to form a grid over the area to be served. They supply water to residential taps and fire hydrants along residential block fronts throughout areas with this occupancy classification. In no case should the pipe size be less than 6 inches in diameter. Larger size pipes may be needed in residential areas for multiple occupancy buildings. In this case, pipe sizing is based on the sum of the peak day water use plus fire-flow requirements. Where there are multiple-occupancy buildings of more than one floor, required pipe sizing is almost always controlled by the fire-flow requirement.

Water distribution piping should be sized and spaced to meet design flow. The minimum size water main for providing fire protection and serving fire hydrants is 6 inches in diameter. Typical values for distribution system piping are summarized in **Table 1-2**. (Reference #8, pg. 12)

Table 1-2

Values Commonly Used in Distribution Piping

Appurtenance	Minimum Standard
Lines	
Smallest pipes in network	6 inch
Smallest branching pipes (dead ends)	8 inch
Largest spacing of 6" grid (8" pipe used beyond this value)	600 ft.
Smallest pipes in high-value district	8 inch
Smallest pipes on principal streets in central district	12 inch
Largest spacing of supply mains or feeders	3,000 ft.
Valves	
Spacing in single and dual main systems:	
Largest spacing on long branches	800 ft.
Largest spacing in high-value district	500 ft.

2. Fire hydrant locations.

All areas served by a water distribution system should have fire hydrants installed in locations and with spacing for fire department use. The following method of locating fire hydrants should be observed in the United States. This method is outlined in Section 614 of the the ISO *Fire Suppression Rating Schedule*–2003 edition. (Reference #9, pg. 12) [Sidebar: Canada uses an area method.]

Briefly summarized, the procedure examines a representative fire-risk location and a computed NFF at that location. The first determination is that a recognized fire hydrant be within 1,000 feet of the fire risk, as fire hose is laid from the fire hydrant to the fire risk. A recognized fire hydrant on a municipal water system must flow a minimum of 250 gpm at 20 psi residual pressure for 2 hours.

The actual flow capability from each fire hydrant in the vicinity of the fire risk is limited by the distance to the fire risk as follows: (Reference #9, pg. 12)

Credit is awarded up to 1,000 gpm from each hydrant within 300 feet of the fire-risk building; 670 gpm from hydrants within 301 to 600 feet of the fire-risk building; and 250 gpm from hydrants within 601 to 1,000 feet of the fire-risk building.

Furthermore, the water utility should review hydrant spacing or representative fire risks in the community with the responsible first-due fire company because the supply hose capacity on fire apparatus may limit the credit assigned by ISO to this item in the *Fire Suppression Rating Schedule*.

The pipe connecting the fire hydrant to the water main is call the **hydrant branch** or **lateral**. Every lateral needs to have an installed valve to enable the water utility to isolate the fire hydrant for repair or general maintenance.

In addition, fire department use typically requires a maximum lineal distance between fire hydrants along street fronts in commercial and congested built-up areas of 300 feet, and 600 feet for light single-family residential areas. Good practice calls for fire hydrants at intersections, in the middle of a block where the NFF equals or exceeds 1,200 gpm, and at the end of dead-end streets.

Summary Statement on Water Supply Distribution

Water supply distribution systems are rather straightforward to evaluate in small communities with a population range up 5,000. The proper evaluation of water supplies and distribution for larger communities, up to cities over 100,000 population, is no simple thing. Water system maps that are kept current and electronic graph records for all of the water supply that enters the distribution system are **essential** to establish an understanding of actual supply versus consumption on an hourly basis, daily basis, monthly basis, and yearly basis. If this information is not in place, the first step to the evaluation of a water system is to put in place a records management system, and then to pay close attention to this system. The water utility has the responsibility to keep public officials, especially fire officials, apprised of the current conditions of the water system. This topic paints the **big picture** for developing and maintaining a comprehensive evaluation of a water system.

References:

1. Mays, Larry W. *Water Distribution Systems Handbook*. New York: McGraw-Hill, 1999, pg. 18.

2. American Water Works Association. *Distribution System Requirements for Fire Protection, AWWA–M-31*. Denver: Author, 1999, pg. 16.

3. *ibid.*, pg. 25.

4. *ibid.*, pg. 24.

5. *ibid.*, pg. 25.

6. *ibid.*, pg. 24.

7. *ibid.*, pg. 19.

8. *ibid.*, pg. 20.

9. Insurance Services Office. *Fire Suppression Rating Schedule*. Jersey City: Author, 2003.

CHAPTER 2: FUNDAMENTAL CONCEPTS OF HYDRAULICS APPLIED TO MUNICIPAL WATER SUPPLY SYSTEMS

OVERVIEW

The design and evaluation of municipal water supply systems is based on both theoretical and applied hydraulics. Hydraulics is the branch of science that defines the mathematical laws of liquids at rest and in motion. This text material is confined to fundamental principles and what is generally referred to as **applied hydraulics**. These fundamentals are essential for understanding many of the considerations involved in the design of a municipal water supply system, the periodic testing of water systems, and the proper evaluation of water systems to assess a given community's water supply with respect to providing adequate water supplies.

A municipal water supply system has the objective of providing an adequate and reliable water supply to meet the following demands:

- residential occupancy water consumption;
- commercial occupancy water consumption;
- industrial occupancy consumption;
- municipal and educational building use;
- Needed Fire Flows (NFFs) that are available from a planned location of fire hydrants throughout the municipality; and
- water for special community needs that include parks and recreation, street cleaning, decorative water fountains, sale of water to contractors through metered water from fire hydrants, etc.

The primary objective of the following material is to present the fundamental concept of hydraulics applied to municipal water systems, in order for municipal officials and fire officials to better understand the design and evaluation of public-sector water delivery systems. Some fundamental hydraulic problems are provided to establish principles used to meet the above objective. A number of tables and charts are provided for future reference by the user of this material in actually working with a specific water supply system.

A supporting objective is to stimulate a more accurate and clearer perception of field practices in evaluating water systems. These practices and procedures need to be based on a fundamental understanding of basic hydraulics. Finally, there are a number of computer-based programs available for the design and evaluation of water pipe systems. One has to understand the fundamentals of water flow in pipe networks to assess whether the computer output is providing reasonably accurate results and to use the computer output to identify significant problems with a water system.

Accomplishment of these objectives requires that operational definitions, concepts, and procedures associated with the science of fluid mechanics and hydraulic theory be related carefully to the subject area of water supply hydraulics. The topics presented below are not unduly difficult, and only a basic understand of algebra is needed to feel comfortable with the level of mathematics presented. Interpretations of the fundamental laws of fluid mechanics are placed in the context of familiar water supply applications so that the reader can quickly establish the relevance of the topics to practical water delivery system concerns.

OPERATIONAL DEFINITIONS

The following definitions provide meaning to terms used throughout the entire text. Other special terms applicable to specific topics will be defined in the subsequent subject matter.

1. Fluid mechanics is the term that defines the physical behavior of fluid systems and the physical laws that describe this behavior. (Reference #1, pg. 46)

    It has broad application in water system design and analysis, and encompasses the behavior of compressible and incompressible fluids, with the primary attention given to water as the fluid under consideration. Such fluids include plain water, water mixed with one or more additives to produce low-expansion foam for both Class A and Class B fires, high-expansion foam, aqueous-film-forming foam (AFFF), carbon dioxide, Halon® and clean-agent extinguishing fluids, and an array of synthetic agents. However, this text material concentrates on municipal water systems and therefore **water** is the only fluid medium discussed in this text.

2. Liquids are fluids that have a definite volume independent of the shape of the container. (Reference #2, pg. 46)

    In conditions of constant temperature and pressure, a liquid will assume the shape of its container and fill a portion equal to the liquid in volume. In most conditions, liquids are considered noncompressible; that is, their volume does not change appreciably under pressure, or with change in temperature. A liquid exposed to atmospheric pressure has a free-standing surface, which means that it seeks its surface level, providing a constant surface datum plane or reference line for calculation. This applies to water storage in a gravity tank, standpipe tank, or other water-holding container. For practical reasons, when the liquid is water, calculations are performed using the **assumption** of a temperature of 70 °F (21 °C) unless there is a sound engineering reason to use a higher or lower temperature because of local climatic conditions.

3. Atmospheric pressure is created by the weight of the atmosphere on the earth. (Reference #3, pg. 46)

   Municipal water systems definitely are affected by atmospheric pressure in relationship to the elevation above sea level. Water contained in pressure tanks on some water systems is an exception. At sea level, atmospheric pressure is 14.7 psi, or 29.9 inches of mercury (Hg), commonly referred to as **one** atmosphere. Atmospheric pressure diminishes with elevation above sea level in accordance with the values depicted in **Table 2-1.**

Table 2-1

Atmospheric Pressure and Elevation

Elevation Above Sea Level	Atmospheric Pressure	
Feet	Psi	Hg (in.)
0	14.7	29.9
1,000	14.2	29.9
2,000	13.7	28.9
3,000	13.2	27.9
4,000	12.7	25.8
5,000	12.2	24.8
6,000	11.8	24.0
7,000	11.3	23.0
8,000	10.9	22.2
9,000	10.5	21.4
10,000	10.1	20.6

4. Hydraulics: The science of hydraulics defines the differential in water head (feet) or the mechanical principles that contribute to placing water at rest or water in motion. (Reference #4, pg. 46)

5. Hydrostatics is the science of water at rest. (Reference #5, pg. 46)

   The scientific laws of hydrostatics defines the principles of water at rest. An excellent example of a hydrostatic condition on a water supply system is where a water storage tank is connected to a water supply pipe using a control valve to be opened on demand. The weight of the water in the storage tank causes pressure on the control valve. The calculation of the pressure on the tank side of the control valve is a hydrostatic problem.

6. Hydrokinetics is the science of water in motion. (Reference #6, pg. 46)

   When a fire hydrant is opened with a cap removed from a discharge spout, the flowing water is a hydrodynamics problem. Hydrodynamics is a general term associated with the science of forces exerted on the pipe wall when water is flowing through a pipe, or the flowing pressure when a quantity of water is discharging from an open pipe or from an orifice device connected to the pipe, such as a fire hydrant opening. The potential energy (i.e., static pressure) becomes kinetic energy (i.e., residual pressure). The weight of the water based on the elevation head supply at the fire hydrant causes the water to move through the piping system and out of the fire hydrant discharge opening. Calculating the water supply and the pipe to supply fire hydrants is also a problem of hydrokinetics.

7. Hydrodynamics is a general term associated with the science of forces exerted by water in motion. (Reference #7, pg. 46) The term generally is applied to hydraulic considerations of fire pumps, but also connotes problems involving the relation of flow, pressure head, and velocity head for closed systems.

8. Absolute pressure (pounds per square inch absolute (psia)): Absolute pressure is the sum of atmospheric pressure (14.7 pounds per square inch (psi)) and the pressure recorded on a gauge (psig). (Reference #8, pg. 46)

   In other words, atmospheric pressure plus gauge pressure equals absolute pressure. Absolute pressure must be evaluated when a liquid is confined by a pressure vessel, and positive or negative pressure forces are exerted on the surface of the liquid. The most common example of this condition is the use of water storage tanks that are under pressure to supply water to pipe lines without a pump interface. Absolute pressure also can be examined from another perspective. Since atmospheric pressure at sea level is 14.7 psi, it is obvious that a gauge pressure reading of minus 14.7 psi represents no pressure. This condition is called absolute zero or 0 psia: 14.7-14.7=0, and provides a reference from which pressure can be measured.

   The relationship of atmospheric and absolute pressures is presented in **Figure 2-1**.

Figure 2-1

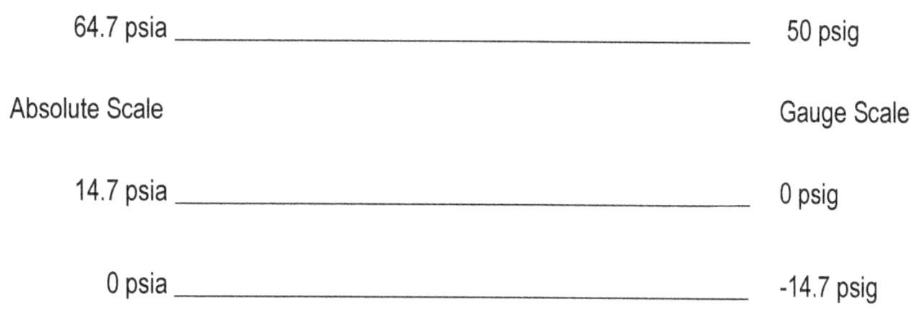

64.7 psia _____	50 psig
Absolute Scale	Gauge Scale
14.7 psia _____	0 psig
0 psia _____	-14.7 psig

It is impossible to make a pressure measurement on the earth's surface unless it is made relative to atmospheric pressure. Therefore, a reference can be established at the atmospheric pressure level, as indicated in **Figure 2-1**. If the pressure one wishes to measure is at the same level, there will be zero pressure relative to atmospheric pressure. Pressure gauges, piezometers, and other pressure-flow measuring devices indicate **gauge** pressure.

Atmospheric pressure is equal to zero pressure on a gauge, often abbreviated as psig on gauges calibrated to register pounds per square inch. Gauge pressures are positive if greater than atmospheric, and negative if less than atmospheric, or measured down from the atmospheric reference. Negative pressure also is called a vacuum.

VACUUM

A perfect vacuum is a space entirely devoid of gas, liquids, and solids. (Reference #9, pg. 46) The literature on this subject indicates that only the National Aeronautics and Space Administration (NASA) in space has ever succeeded in exhausting all the air from a closed vessel. This applies to suction pipes on stationary fire pumps. Therefore, the word **vacuum** actually means **practical vacuum**, and is measured by the amount of this pressure below the prevailing atmospheric pressure. Vacuum is measured by **gauges** graduated in inches of mercury (Hg).

Physical Characteristics of Water

Water is the only fluid considered in this text material. Therefore, it is important to define its physical characteristics and properties that affect hydraulic calculations involving both public and private water supply systems. The following characteristics should be considered for all calculations involving water supply systems.

DATUM LINE

A datum line is a line designating the potential energy of water; also a line that serves as the basis for mathematical computations. (Reference #10, pg. 46)

Water is a colorless liquid that takes the shape of the container in which it is placed. The top surface of the water seeks its own level and, therefore, is flat for practical considerations. The top level of the water in a container is often referred to as the **surface datum line**. The measurement is given in feet above sea level, or some other reference point. This concept is illustrated in **Figure 2-2**. Note that the underground pipe reference line may change with time if water is being used from the holding tank. In hydrokinetic problems, the water surface datum line is often referenced as the potential energy line.

Figure 2-2

Water Service Datum Line

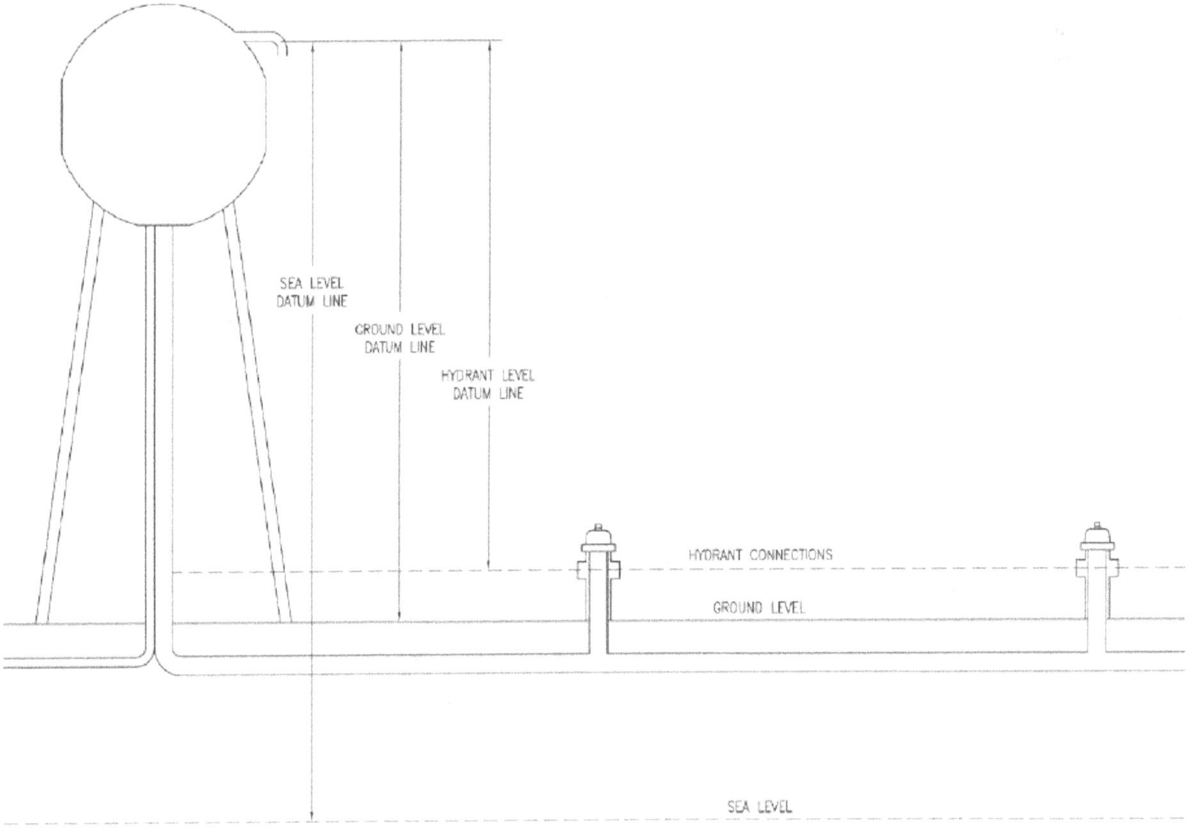

Water has weight, normally expressed in pounds per cubic foot, (lb/ft³) or kilograms per liter, (kg/L). A commonly used value for fresh water is 62.4 lb/ft³. This constant is based on the weight of a cubic foot of water when the water temperature is 50 °F (10 °C), the pressure is atmospheric, and the water has been treated for human consumption. Therefore, it should be recognized that the water varies with temperature, pressure, and impurities or additives.

Figure 2-3

Variation of Weight With Temperature

Temperature (Degrees F)	Weight (lb/ft³)	Temperature (Degrees F)	Weight (lb/ft³)
32	62.416	75	62.261
40	62.423	80	62.217
45	62.419	85	62.169
50	62.408	90	62.118
55	62.390	95	62.061
60	62.198	100	61.998
65	62.000	150	61.203
70	62.000	200	60.135

Variation With Weight and Pressure

Pressure (psi)	Weight (lb/ft³)	Pressure (psi)	Weight (lb/ft³)
0	62.40	20,000	62.82
1,000	62.421	30,000	63.02
10,000	62.81	65,000	68.60

Note: These figures are approximate, as compressibility also varies with pressure.

Variation of Weight With Impurities

Source	Weight (lb/ft³)
Garonne River, France	62.409
Thames River, England	62.419
Mississippi River, U.S. (filtered)	62.415
Springs, West Virginia	62.419 to 62.77
Pacific Ocean	64.05
Dead Sea	73.17

Note: Weight of water variations around the globe.

Figure 2-3 illustrates the effect of these three variables on the weight of a **cubic foot** of water. With the exception of variations due to impurities, the magnitude of variations is negligible. For practical hydraulic calculation of water supply problems, the weight of water is considered to be 62.5 lb/ft³. However, for more precise calculations, this average condition value must be replaced by the actual measured value when experimental work is being conducted.

Fundamental Characteristics of Water as a Fluid Medium

The preceding material has established that liquids in general have a definite volume but no definite shape. It is important to consider the volume of water in different shaped containers that have equal volumes.

EQUAL VOLUMES

Figure 2-4 shows a tank with a volume of 1 cubic foot (ft³) that measures 1 linear foot on each surface edge. When the container is filled, it will hold 1 ft³ of water. In contrast, the cylinder shown has a diameter of 13-9/16 inches, giving a computed cross-sectional area of one square foot (ft²). Now consider that the liquid is poured from the cube into the cylinder.

Figure 2-4

Cubical and Cylindrical Tank Volume Comparison

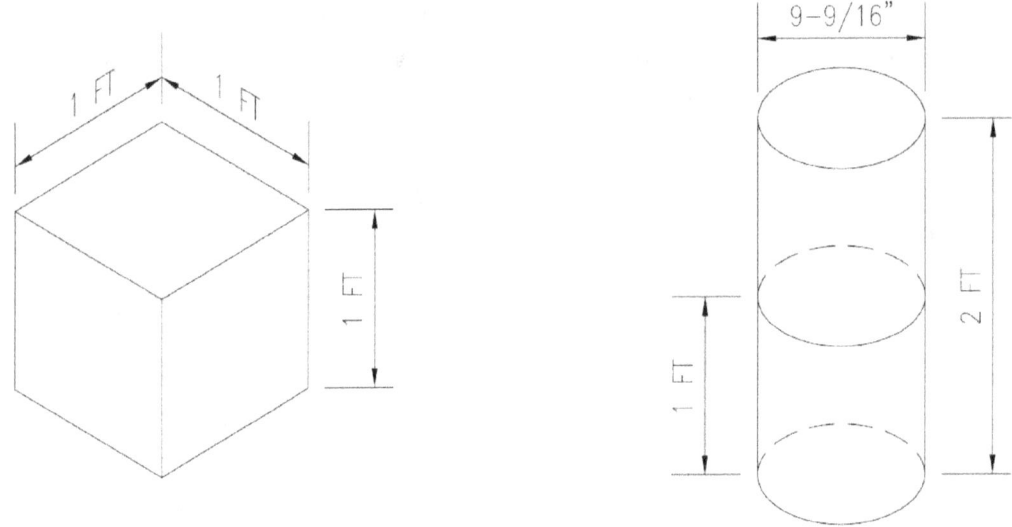

How high will the liquid rise in the cylindrical container? This can be determined from Formula 1:

Formula 1   Volume=Base Area x Height

Height=Volume/Base Area=1 ft³/1ft²=1 ft

Therefore, when the base areas are equal, equal volumes of liquid will rise to the same height, even though the containers have dissimilar shapes.

SHEAR STRESS AND VOLUME

Shear stress is the reaction to tangential forces applied in opposite directions to an object, a medium, or a container, at the top and bottom surfaces. (Reference #11, pg. 46) The water tank in **Figure 2-5** illustrates the principle of shear stress applied to equal volumes.

Figure 2-5

Shear Stress Applied to Volume

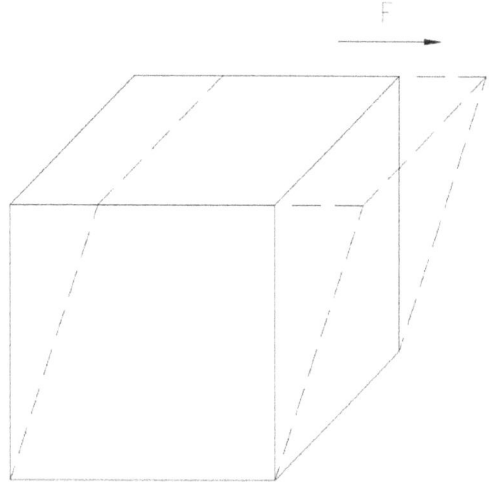

The formula is:

Formula 2 $S = F/(L x W) = F/Area$

Where:

S = Shear stress

L = Length

W = Width

Force = Pressure x Area

Experiments have shown that the 1-cubic-foot container cannot sustain the opposite forces in equilibrium. (Reference #12, pg. 46) If it could, it would be possible to let out the contents of the tank on a street and the liquid would retain its shape. The fact that the liquid will spread quickly over the street and seek a common level is evidence that liquids, including water and water additives, cannot sustain a shear stress. Note that the volume has not changed; all that has happened is that the shape of the liquid has changed. Calculation of shear stress is based on fluid principles explained below.

UNIFORMLY DISTRIBUTED PRESSURE

The concept of pressure being distributed uniformly over a surface area must be examined for definition of the intensity of pressure, or unit pressure. Refer to **Figure 2-6**. Since the depth of the liquid in the container has an equal dimension or value over the entire area on the bottom of the tank, the intensity of the pressure is the same at every point on the bottom surface, and the pressure is uniformly distributed. This fact provides a general principle for establishing an equation for **total pressure**.

Figure 2-6

Intensity of Pressure on Surface Area

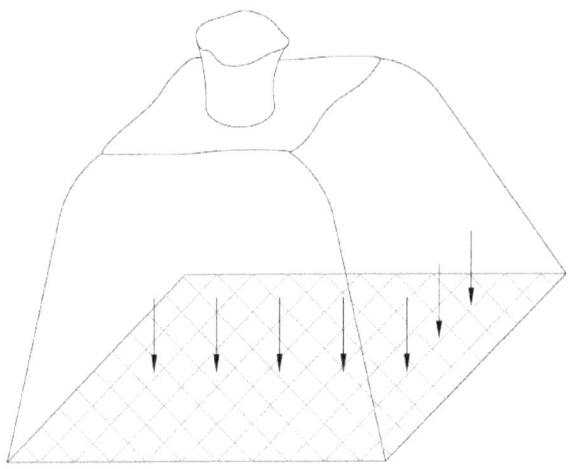

Total Pressure: When the pressure caused by a force is distributed uniformly over a plane surface, the **total pressure** on the surface is equal to the product of the intensity of pressure and the area of the surface. (Reference #13, pg. 46) This leads to Formula 3:

Formula 3 $F = p \times A$

Where:

F = total force on the surface

p = intensity of the pressure

A = area of the surface

For large areas, such as the base plane in water-holding tanks or reservoirs, the intensity of pressure p is usually expressed in pounds per square foot, abbreviated as psf, and the area, A, subjected to this pressure is then expressed in square feet. However, for hydraulic calculations associated with water systems, pressure normally is considered as being distributed over comparatively small areas. Consequently, the practice has developed for expressing the intensity of pressure p in psi (Pascals in the International Standard of Units (SI) system.)

The formula and relationship of pressure intensity provides a concept of the distributive reaction of water flowing in pipes, and this leads to the fundamental question, what is pressure? The most universal and acceptable definition of **pressure** is stated as follows: Pressure is defined as a force per unit area when the force acts at right angles to a surface. (Reference #14, pg. 46)

PRESSURE AS A RATIO

It follows from this that pressure is the ratio of the perpendicular force, acting against any surface, to the area of that surface.

Pressure = Force/Area (over which the force acts)

In the basic language of water supply hydraulics, the terms **force** and **pressure** often are used loosely to mean about the same thing, but in hydraulic theory they mean two different fluid functions. Pressure is the amount of force applied to the defined area over which it acts. It is measured in such units as psi, grams (gm/cm^2), dynes ($dynes/cm^2$), or Pascals. For example in municipal water systems, it is appropriate to speak of pressure in an underground main as being so many psi.

Formula 4 $F = PA$

Pressure, therefore, can be considered as the amount of force concentrated on a small area, such as a square inch, or square centimeter. Force is the amount of pressure over the entire surface area of the fluid.

The concept of pressure often gives more information than the concept of force, as a simple example will show. If an object weighing 5 pounds and having a large flat base is held in your hand (**Figure 2-7**), the weight (force) is distributed evenly over a large area and no part of the hand is unduly compressed. If the object is inverted so that smaller end rests on the hand, the surface of the hand under this end will be depressed much more than in the broader distribution. The pressure is greater because the total downward force (pressure) is the same, but the area of contact is less. (Reference #15, pg. 46)

Figure 2-7

Examples of Force Distribution

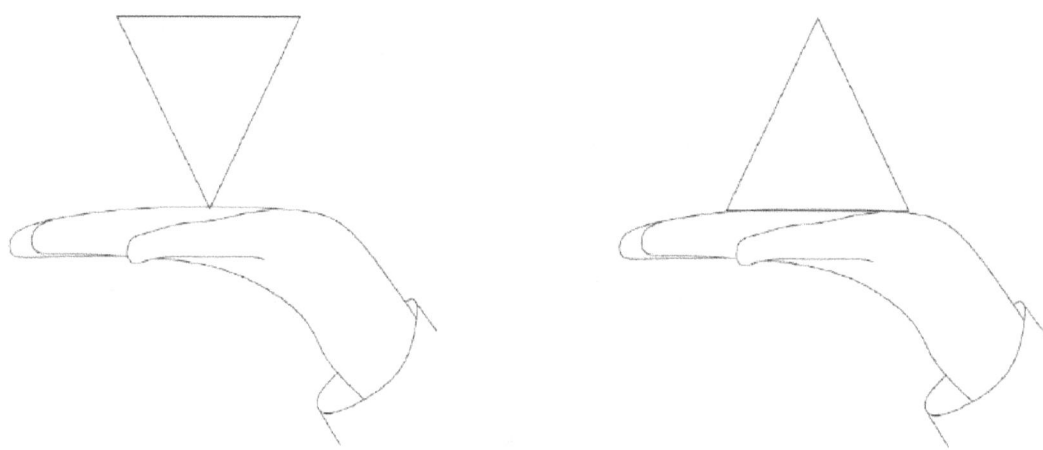

This concept may be applied to a water supply situation. Refer to the two float valves in **Figure 2-8** used on the intake side and the discharge side of a small gravity tank used as a priming tank for a stationary water supply pump taking water under lift as from a river. This illustrates the fact that the pressure function is the same on both valves A and B, but the force on A is greater than B.

Figure 2-8

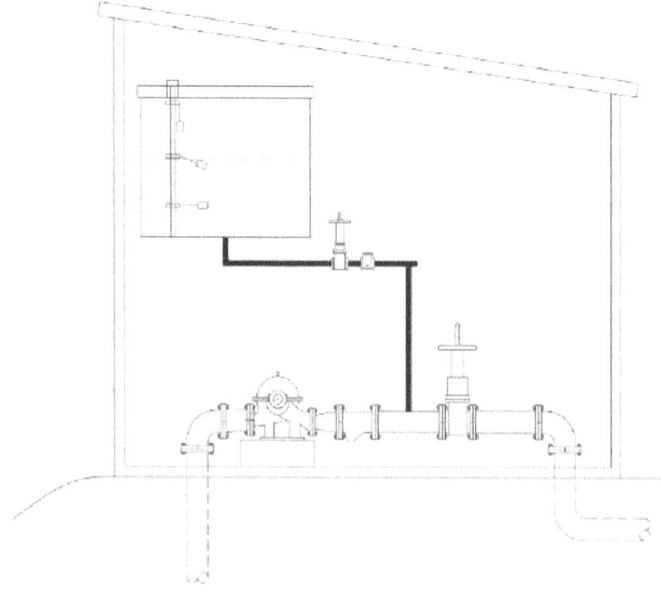

Next, refer to **Figure 2-9**. Assume that the riser pipe on the gravity tank to a small municipal water system is 100 feet high and consists of 6-inch pipe (left-hand illustration).

Figure 2-9

Gravity Tank Illustrations

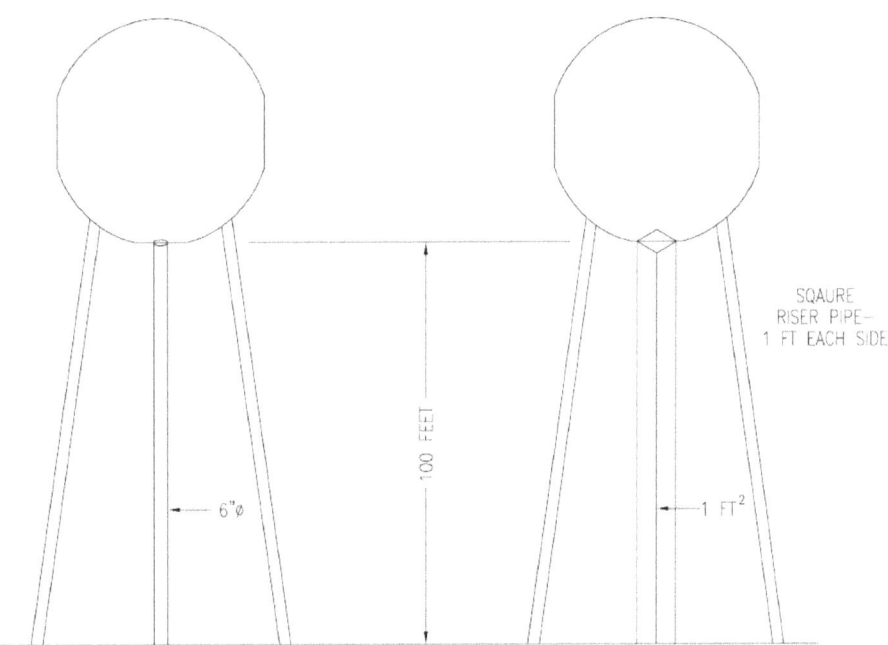

The volume of the riser pipe, or the filled capacity of the pipe, can be determined from Formulas 6 and 7,

Formula 6   Area (Pipe Base) = $hD^2/4$ or $3.1416D^2/4$

Formula 7   Volume = $0.7854 D^2h$

Where:

D = diameter and h = height

Problem 1

What is the intensity of pressure in psi at the base of the riser pipe?

Solution:

1. The total pressure in the riser pipe is a function of the total weight of water contained in the pipe.

2. Therefore, the volume of water in the riser pipe equals

$V = 0.785 D^2h$

$V(ft^3) = 0.7854 (.5^2 \times 100 \ ft)$

$V (ft^3) = 19.636 \ ft^3$

Note: All units of measure are in feet.

3. Each cubic foot of water in the riser pipe weighs 62.5 lb/ft³.
4. Therefore, the total pressure in lb/ft³ is calculated by:
   a. 62.4 lb/ft³x19.635 ft²=1,225 pounds
   b. 1,225 pounds divided by (0.7854x0.25)=6,239 lb/ft²

An important point to observe is that the verbal expression, pounds per cubic foot, means total pounds dived by cubic feet of volume, or pounds/volume. Keeping units consistent, particularly in the English system of mathematics, is one of the difficult tasks in solving fluid mechanics and hydraulic problems.

5. The total pressure may be expressed in psi as follows:
   a. The weight of the water in the riser pipe remains the same as calculated above: 1,225 pounds.
   b. However, the weight is distributed over a square inch area as follows:
      $A=0.7854d^2=0.7854x6^2=28.27$ in²
   c. Therefore, 1,225 pounds divided by 28.27 in²=43.3 psi

Referring to the gravity tank riser Problem No. 1, a determination was made that the total weight of water (W (lb) = r A x h). The pressure at the bottom of the riser pipe surface, the distributed reaction point, will be the total weight of the water divided by the surface area, or P=W/A=r A h/A=rh/unit area. As illustrated in the problem, if one is working in pounds and feet, the unit of P will be pounds per square foot (lb/ft²). If one is working in inches and pounds, the unit p will be pounds per square inch (psi).

Next, suppose the fluid being used for fire suppression is changed from fresh water to salt water. This change of weight per cubic measure brings into consideration a factor termed specific gravity (Sg). The term is defined below.

SPECIFIC GRAVITY

Specific gravity is the ratio of the weight of a given volume of a substance to the weight of an equal volume of water. (Reference #17, pg. 46) This ratio may be quite important in making hydraulic calculations for water densities other than fresh water, or the mixture of additives in the water used for fire suppression operations and fixed systems. This concept is extremely important for the future of municipal water system. Cities in the United States are using a desalination process to convert salt water to potable water for municipal water systems. Tampa, Florida, represents one of the pioneering efforts in using salt water for domestic water systems. (Reference #16, pg. 46) Two problems illustrate this point.

Problem 2

The specific weight of salt water is 64.0 per cubic foot for calculation purposes. Determine the specific gravity (Sg) of the salt water for a stationary municipal hydraulic water pump.

Formula 8   (Sg) =   Specific Weight of the Substance

                                   Specific Weight - Equal Volume of Water

(Sg)=64.0 lb/ft³/62.4 lb/ft³=1.03 for salt water

Problem 3

The specific gravity of a given commercial brand of mechanical foam used for fire suppression is 0.685. Determine the specific weight (f) of the foam. Knowing the Sg of the foam, the relationship rf=rw x Sg can be used to calculate the answer as follows:

rf = 62.4 lb/ft³x0.685=42.7 lb/ft³

Pressure Principles

The previous discussion has focused on establishing some fundamental concepts and how they have relevance to water supply system hydraulics. These concepts now permit introduction of several statements that are generally described in the literature as Pascal's Laws. His principles are as follows: (Reference #18, pg. 46)

Principle 1: Pressure at any depth beneath the surface of a given liquid is directly proportional to the depth. The pressure of a liquid on a horizontal surface is numerically equal to the weight of the column of liquid of uniform cross-section directly above the unit area of the surface. The right-hand illustration in **Figure 2-9** shows the gravity tank redrawn to provide a **square** riser pipe instead of the 6-inch-diameter riser pipe originally referenced.

The right-hand gravity tank in **Figure 2-9** has been modified to add a dimensional factor. The riser pipe has been partitioned into modules measuring 1 foot on each side. Remember that for normal hydraulic calculations, a cubic foot of water weighs 62.4 pounds for practical purposes. Now, assuming that the water is contained in a cubic vessel measuring 1 foot on each side, the total weight of 62.4 pounds is exerted on the bottom area of 1 square foot or 144 square inches. The pressure is then 62.4 pounds per square foot or 62.4 lb/ft² divided by 144 in²/ft²=0.433 psi. Each 1-foot increment on the riser pipe must be multiplied by the determined pressure for 1 foot of height. The total weight of the water is doubled at the 2-foot increment level, and as the pressure function is increased any number of feet, h, the pressure on the bottom increases h times. For any depth of water, the pressure is calculated by P (lb/ft²) = 62.4h, where h is in feet and the units are pounds per square foot; or P (lb/in²) when the unit of measure is psi. This principle permits calculation of a constant for computing pressure in psi from a given or assumed head differential.

The calculation is accomplished as follows:

Formula 9 $P(psi) = \dfrac{62.4 \text{ lb/ft}^3}{144 \text{ in}^2/\text{ft}^2} = 0.433h \text{ (ft) for water}$

When the specific gravity of the liquid changes, the basic formula can be adapted to represent the change from which has a specific gravity of 1.0 in the following formula:

Formula 10 $P = 0.433 \times Sg \times h$

Where h is in feet, and Sg represents the specific gravity of the liquid other than water.

In the discussion of intensity of pressure, it was determined that a 100-foot-high riser pipe would create a pressure of 43.4 psi at the base of the riser. Using the same dimension for the height of the riser pipe, **Formula 10** now can be adapted to provide the same answer as follows:

Problem 4

If a 6-inch riser extending 100 feet high on a gravity tank is full of water, what is the pressure in psi at the base of the riser?

$P = 0.433 \times Sg \times h$ where water has a specific gravity of 1.0. Substituting:

$P = 0.433 \times 1.0 \times 100 \text{ ft} = 43.3 \text{ psi}$

Since this text material is limited to the use of potable water in municipal water systems, **Formula 10** can be modified to drop out the specific gravity function, giving **Formula 11** for use with municipal water systems that use potable water.

Formula 11 $P(psi) = 0.433 \times h(ft)$

DETERMINING HEIGHT FROM A KNOWN PRESSURE

A converse relationship may be established between pressure and potential head. The illustration and computations with a gravity tank only consider a water column the height of the riser pipe; this assumption neglects potential water in the gravity tank. It may be desirable to determine the height of the water in the gravity tank from the recorded pressure at the base of the tank. **Formula 11** may be solved for h in the following manner.

The equation P=0.433h has been established. This equation maybe used to solve for h in feet. Therefore:

$$h(ft)=P/0.433$$

If P=1 psi, the formula can be written as:

Formula 12 $h(ft)=1 \text{ psi}/.433$ or $h(ft)=2.31 \text{ P}$

Figure 2-10

Partitioned Riser with an Inspector's Test Gauge at the Base of the Riser Pipe

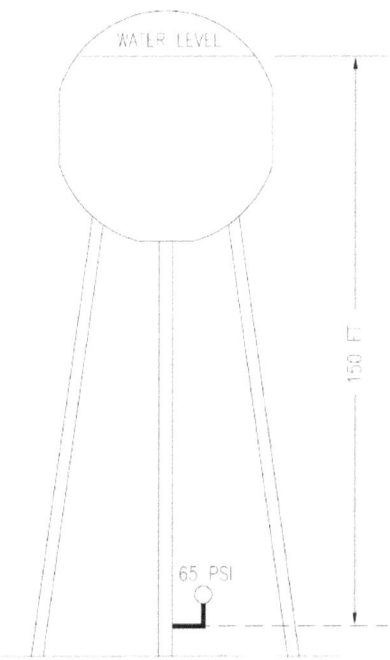

Problem 5

Figure 2-10 shows a pressure gauge attached to an inspector's test port at the base of a gravity tank. The gauge reads 65 psi. What is the height of the water in the gravity tank above the gauge tap?

Given, P = 65 psi

Therefore: **Formula 12** applies using the following substitutions:

h = 2.31(psi/ft)x65 psi

h = 150 ft

In summary, the water level stands 150 feet above the inspectors' test tap and creates a static pressure of 65 psi on the gauge. A water tank generally is filled to the overflow point. By knowing the elevation of the overflow point above the gauge tap location, one can determine whether a water tank is full or only partially full. By knowing the height of the riser pipe to the bottom of the tank, one can determine just how full that water tank is at any given time.

The previous discussion on hydraulic principles relates only to hydrostatic conditions. The computations apply only to water at rest; there is no flow of water in the storage tanks or pipes under consideration.

Principle 2: Pressure at any point in a liquid (i.e., water) is independent of the directional force. It may seem difficult to understand that, at any depth beneath the surface of water, there is a force upward due to pressure equal to the force downward. If this were not true, an infinite series of unbalanced forces would develop; any volume of the water would be under the action of unbalanced forces that are greater in one direction than another, and a perpetually circulating current would exist. Indeed, the various parts of the water would not remain at rest, and a condition of equilibrium could not exist.

A pressure function is not considered to have direction, but the force which results from it and which acts on any given area does have direction, which is the line perpendicular to the area. Hence, one must say that at any given point in the volume of water, pressure is independent of the direction, whereas at a surface it produced a force perpendicular to that surface. This concept is illustrated in **Figure 2-11** by the use of a fire hydrant with a large diameter and two small diameter outlets. Cap type Bourdon gauges are attached to each outlet, and the hydrant is opened. It is observed that the pressure reading is equal on all three gauges. This demonstrates that pressure is independent of direction at a common level of entrapped water.

Figure 2-11

Pressure Functions on Fire Hydrant Outlets

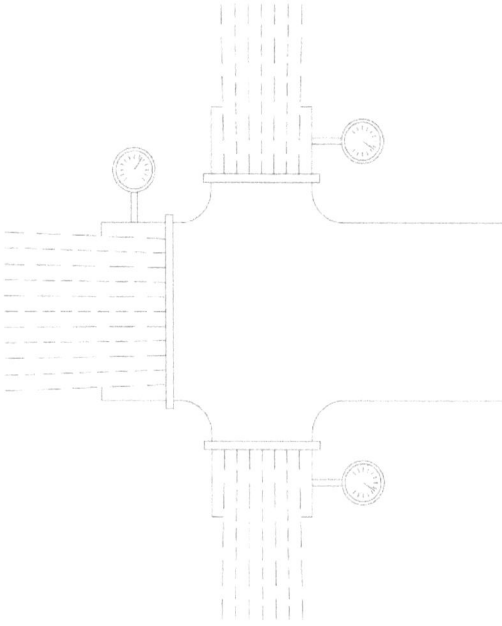

Remember, the pressure intensity in psi is a function of the potential head differential on the discharge outlets to the fire hydrant. Furthermore, the conditions under consideration are still **static** (no flow) in nature.

Principle 3: Pressure applied to a fluid is transmitted undiminished throughout the fluid. (Reference #19, pg. 46) A clear understanding of this principle is fundamental to the computation of transmission of pressure forces in underground water main distribution systems. Assume that a primary feed main on a water system supplies two-directional water flow to secondary water mains as illustrated in **Figure 2-12**.

Let it be further assumed that pressure gauges are placed on the secondary water mains at fire hydrant equidistance (i.e., 500 feet apart), with no water flowing thought the distribution system. Each gauge will show the same pressure at the same distance from the tee connection to the primary feeder.

A basic understanding of this pressure principle is essential for the proper calculation of water supply problems where the supply source splits from a common junction point to supply discharge devices, such as fire hydrants, hose nozzles, and sprinkler heads on automatic sprinkler systems.

Principle 4: The downward pressure of a fluid on the bottom of a vessel is independent of the shape of the vessel. (Reference #20, pg. 46)

Again water gravity tanks represent a situation where the shape and size of the holding vessel do not have standard configuration. **Figure 2-13** shows five typical water storage tanks, all with different holding tank designs. However, the water level in each tank above ground level is the same: 150 feet. Due to the counterbalancing of forces in the water contained in each water tank, the pressure in psi will be the same at the tank base. Using **Formula 12**, it can be determined that P (psi)=0.433x150 ft=65 psi at the base of each tank.

Figure 2-12

Primary Water Mains Feeding Secondary Water Mains

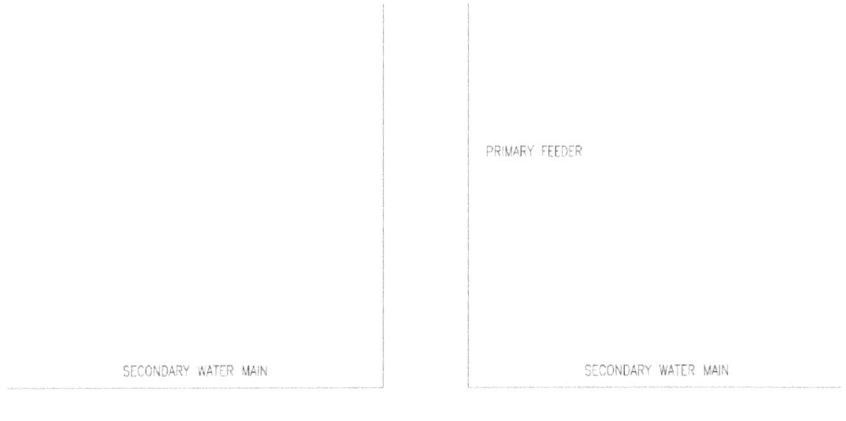

PRIMARY FEEDER

SECONDARY WATER MAIN   SECONDARY WATER MAIN

Figure 2-13

Downward Pressure on the Bottom of Vessels

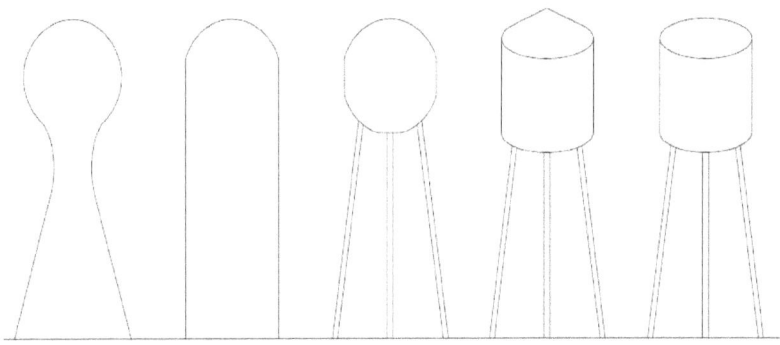

Typical Water Storage Tanks

A SUMMARY AND EXPANSION OF FIRST PRINCIPLES OF WATER PRESSURE

The term hydrokinetics was defined earlier as a division of the broader science of hydromechanics. The discussion to this point has been confined essentially to the application of water at rest in which the water weight was the only property of significance. It also is of importance to consider principles of water pressure in a motionless state (static) and in motion (dynamic). These new principles follow.

Pressure Head

The height of water is given a special designation, pressure head. Pressure head represents the vertical height, in feet or inches, which causes a pressure intensity at the bottom of a column of water. (Reference #21, pg. 46) For every pressure intensity level, there is a corollary pressure head. In a massive body of water, such as an impounding reservoir or municipal water supply system with a large standpipe or gravity tank, the measured depth of water and pressure head are synonymous; that is, the pressure intensity varies with the depth. For a water distribution piping system, the pressure head with no water flowing in the network is equal to the height to which the column of liquid could be raised. This is known as the hydrostatic condition, or simply **static pressure**.

Note the difference in units between pressure (psi) and pressure head (feet).

Static Pressure

Static pressure is pressure intensity created by a confined liquid at rest. (Reference #22, pg. 47) Static pressure may be developed through the difference in elevation between two points of a confined liquid, or it can be created artificially by pressure tanks or different types of mechanical pumps. In gravity feed systems, static (elevation) pressure is the pressure in psi corresponding to the vertical distance in feet between the elevated point of supply (the bottom of a gravity tank, a pump suction reservoir elevated location where city fire flow tests are made) and the level of water delivery. Static pressure is calculated between discharge heads at different levels; between heads and junction, or reference points where pressure adjustment must be made; and between heads and the level point of supply. Static pressure has no relation to size and length of pipe.

Potential Energy to Kinetic Energy

Theories concerning potential energy and kinetic energy follow the principles developed by Sir Isaac Newton. (Reference #23, pg. 47) One is related to potential energy, which is energy that exists, or is stored, or otherwise usable. A reservoir that supplies water to a municipal water distribution system has potential energy because its contents flow to a lower level and gain pressure from gravitation.

Figure 2-14

Elevated Gravity Tank Supplying Three Fire Hydrants

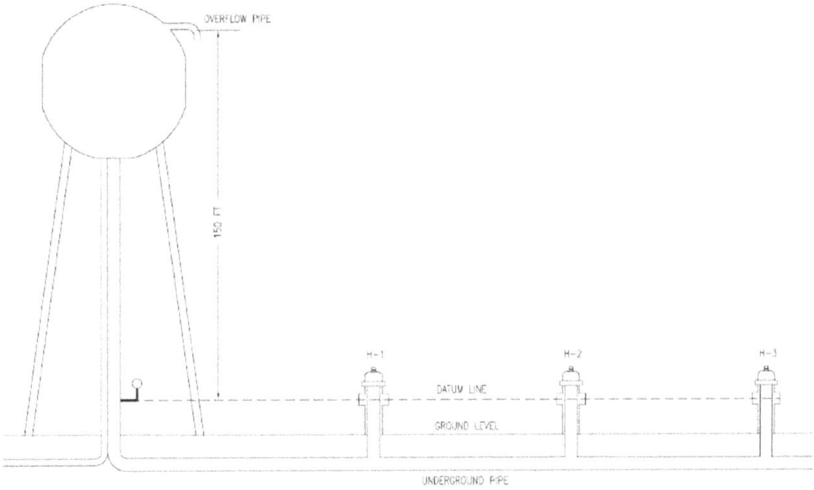

This concept also can be illustrated using a gravity tank as illustrated in **Figure 2-9**. The weight of the water in the gravity tank referred to on page 32 represents potential energy. The intensity of this pressure has been calculated in Problem 5. **Figure 2-14** illustrates that the water gravity tank is now connected to an underground pipeline supplying three fire hydrants. The discharge outlets for each hydrant are positioned at the same level as the inspector's tap on the gravity tank. If a pressure gauge is attached to the outlet of each fire hydrant, the reading would be 65 psi because the recorded potential energy is on the same datum line.

Sea level is used most often in hydraulics as the base datum line or reference line. In **Figure 2-14**, the hydrant outlets are all on the same elevation line (datum) line; this line is also at the same level as the inspector's test tap on the gravity tank riser. Therefore, the discharge reference points in the problem are considered to be on the same datum plane.

KINETIC ENERGY

A second force is **kinetic energy**, which is determined by the motion or velocity of a body. This results from transforming potential energy into a state of motion with a corresponding velocity function. The basic formula for this relationship follows. (Reference #23, pg. 47)

Formula 13 $KE = 0.5 \, M V^2$

Where:

KE = kinetic energy

M = mass of the body

V = velocity in feet per second

In the Newtonian definition, the principle of conservation of energy states that the total energy in the system remains constant. Thus, a change in potential energy presumes a corresponding change in kinetic energy; PE (potential energy) = KE (kinetic energy). Or, as the potential (stored water in the gravity tank) energy passes from the stored state to a condition of doing work (i.e., moving down the riser, throughout the underground pipe, and discharging from hydrant 3) it is converted into kinetic energy (KE). (Reference #24, pg. 47)

The term kinetic energy implies motion; that something is moving. Water flow may be steady, erratic, uniform, or nonuniform. The flow may or may not have uniform pressure and density. In this text material applied to municipal water system, it is assumed that the density of the water will remain constant unless otherwise specified.

Steady flow exists when, at a particular point in a flow system, the velocity, pressure, and density of the water remain constant with respect to time. This represents the common type of flow in most types of municipal water supply problems that require hydraulic evaluations. One example is the flow in a water main when the quantity of discharge from a fire hydrant is constant. The main does not need to have a uniform cross-sectional area.

Unsteady flow exists at a point in a water system when the velocity and pressure of the water change with respect to time. An example is the water behavior in a water main system when the flow, or rate of discharge from the main, is varied by opening or closing a discharge valve. Therefore, the type of conditions to be considered include

💧 Uniform flow when the cross-sectional area is constant, such as in a fire hydrant flowing full.

💧 Nonuniform flow, when the outlets are changed. This happens when a fire hydrant outlet has a cavity due to the lack of pressure to fill the entire orifice area.

Kinetic energy must be considered in the calculation of flow characteristics in water distribution system, in fire protection mains, automatic sprinkler systems, and related fire-suppression equipment including hoselines. Calculations require a fundamental understanding of flow velocity, the continuity of flow in pipe systems, and the relationship between pressure functions at defined reference points in the flowing network. Torricelli's equation, Bernoulli's equation, and the Equation of Continuity provide a foundation for establishing flow principles and defining terms.

VELOCITY FLOW OF WATER FROM A GRAVITY TANK

Once again, **Figure 2-15**, illustrating a gravity tank supplying a simple underground piping supplying fire hydrants, can be used to establish relationships between flow, velocity, and pressure. A first concern is the importance and impact of velocity flow in pipe networks and the velocity emitted from an orifice, normally called the velocity of **efflux**.

Figure 2-15

Example of Pressure Loss in Water System Flowing Water

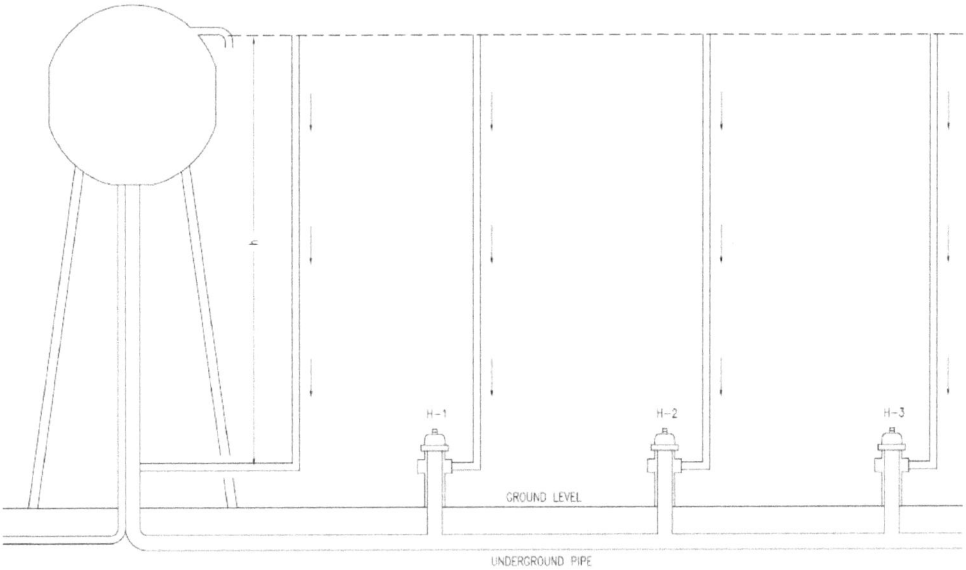

When water flows from a gravity tank through fire hydrant 3, the velocity of **efflux** is greater when the height of the water above the opening is greater, all other factors being constant. According to Torricelli's theorem, the velocity of efflux **is not** proportional to the square root of the height; it is equal to the velocity of a body falling freely from rest over a distance equal to the depth of the orifice below the free surface of the liquid. To double the velocity of efflux from such an orifice, one must quadruple the height of the water. (Reference #25, pg. 47)

Torricelli reasoned as follows. The potential energy of a mass of water, M at the hydrant opening (3), due to the pressure above it, is the same as the potential energy: mass x gravity acceleration x potential height (i.e., height of water above the hydrant opening). When the water flows out of the tank through the riser, and the underground pipe and out the hydrant opening, the potential energy is converted into kinetic energy. The following formula expresses this condition: (Reference #26, pg. 47)

Formula 14 $V \text{ (velocity of efflux)} = (2gh)^{1/2}$

For purposes of illustration, a freestanding tube is inserted into the side of the tank in **Figure 2-15** at the water line and extended horizontally on the same datum plane to a location over hydrant 2. The significance of this concept is summarized in the following statement.

With reference to **Figure 2-15** water flows from the gravity tank filled to a depth h with the same speed as it would acquire by free fall through the same distance. The comparison assumes no friction at the opening. However, on account of frictional resistance in the pipe network from the hydrant back to the gravity tank, the actual velocity would be less than the theoretical velocity.

THE FUNDAMENTAL FLOW EQUATION

Based upon the early works of Torricelli and Chezy concerning the velocity of flow through a closed conduit, a determination was made that the quantity of water passing any defined point in a pipe section is a function of the area of the pipe (assuming the pipe is flowing full and the pipe perimeter is completely wetted) and the velocity flow. (Reference #27, pg. 47) Although the general reference to measuring quantity flow at the point of discharge (an orifice opening) is common in water system hydraulics, it is important to establish that the flow may be calculated in an underground pipe or in a section of vertical pipe or overhead mounted pipe, from the known values of area and velocity. The kinetic energy causing flow at a defined point in a water system is calculated by the following formula, often referenced as Torricelli's Theorem. (Reference #28, pg. 47)

Formula 15 $Q=AV$

Where:

Q=cubic feet per second

A=area of the pipe or orifice in feet

V=the theoretical velocity in feet per second

Note: All measurements are in feet

EQUATION OF CONTINUITY

One concept which must be satisfied in all water flow problems is continuity of flow. This recognizes that no water is lost or gained and no cavities are formed or destroyed as the water passes through a pipe conduit. A proper understanding of flow continuity is essential to the design and evaluation of water supply systems. When the fluid is essentially noncompressible, such as water, continuity is expressed in the following equation: (Reference #29, pg. 47)

$Q=A^1V^1=A^2V^2=A^3V^3$

Figure 2-16

Water Supply System with Attached Gauges

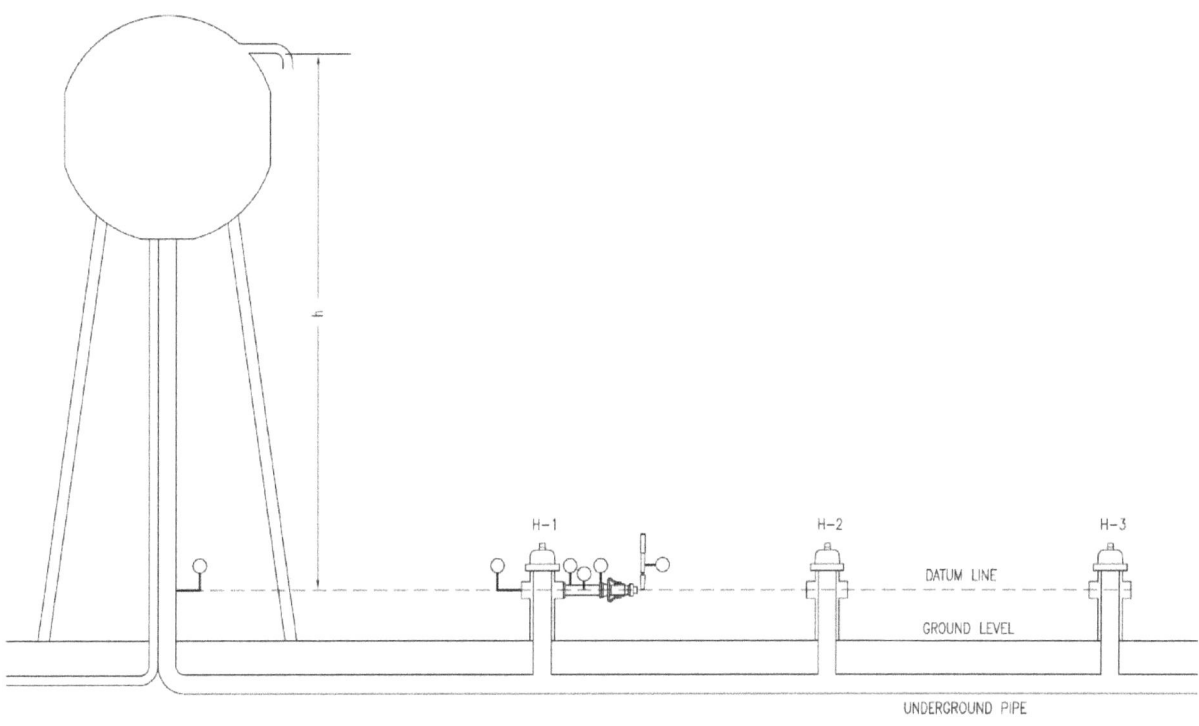

The concept of flow continuity can be examined and analyzed in relation to the basic gravity tank water supply system shown in **Figure 2-16**. This is a simple illustration showing that the gravity tank on a water system supplying fire hydrants also is supplying a standpipe with hose connection outlets in a multistory building for fire protection. In a municipal water system the standpipe actually would be supplying a number of underground mains for both domestic and fire protection usage. Essentially this is an extension of **Figure 2-15** that simply illustrates supply to fire hydrants. In this specific case, assume that a hoseline is connected to the upper-level floor, and that the discharge is 150 gpm. This is approximately 20 ft^3/min. The underground pipe has a cross-sectional area of 0.196 ft^2 for a 6-inch water main. The velocity of flow in the pipe is 20/0.196 = 102 feet per min or 1.7 ft/second. The water flow past fire hydrants 1, 2, and 3 is equal. This also represents the same quantity of water flowing from the hoseline in the multistory building. However, it is important to note that the continuity equation does not identify the energy dissipation in the pipe section from the gravity tank to the standpipe hose outlet in the illustrated building. This energy dissipation is commonly called **friction loss,** a topic that will be examined in detail in relation to the kinetic energy function in the water supply system.

THE BERNOULLI THEOREM

The work of John Bernoulli in 1772 was a continuing exposition on the theory of water in conduits. (Reference #30, pg. 47) The basic principle that Bernoulli discovered is the one most often used in hydraulics and is generally described as the law of conservation of water energy, or simply Bernoulli's Theorem. It is one of the most fundamental and far-reaching statements concerning fluid mechanics and it applies Newton's law of conservation of energy to the flow of water. It is stated in the equation:

$$V_1^2/2g+P_1/w+Z_1=V_2^2/2g+P_2/w+Z_2$$

The values in this equation are defined as follows: V is the velocity in ft/sec; g is the acceleration of gravity which is 32.2 ft/sec^2; P is pressure in lb/ft^2; Z is elevation head in feet; and w is the specific weight of the fluid in lb/ft^3. Essentially, the equation states that the total head is equal to the pressure head plus the velocity head plus the static head at reference point 1; and this relationship holds at all other points in a continuous fluid medium. It stands to reason that this includes water. Since all terms in the above equation are given in units of head (feet), each term in the expression is called respectively, velocity head, pressure head, and potential head. Elevation head and pressure head are related directly to the potential energy of a given water system. The potential energy may be considered as the difference in height or head through which the quantity of water can flow from an established elevation level to one or more defined flow points. **Figure 2-17** has been modified to illustrate the potential energy levels, or the Z potential concept at the end of the three fire hydrants and the top hose outlet on the building standpipe. This concept is basic to other applications of hydraulic theory applied to fire protection problems.

As a first concept, consider the relationship of potential head for the first two fire hydrants between the gravity tank and the warehouse. Since $Z_1 = Z_2$ because they are at the same elevation level or datum line, the two terms cancel out of the analysis of flow at each hydrant reference point. Therefore, the Bernoulli Equation can be written to express flow energy with reference to hydrants 1 and 2; the basic equation now can be referenced as the following formula:

Formula 18 $V_1^2/2g=V_2^2/2g+H(L)$

Where: H(L) is treated as energy loss to the resistance of a given quantity of water flow between the two fire hydrants. Head loss may be defined broadly as friction loss, although friction loss is influenced by a number of factors. In summary, it should be noted that at hydrant 1, head pressures, $H(p)=V_1^2/2g$ (velocity head as a function of kinetic energy) + the pressure intensity, P/w. Therefore the total energy at hydrant 1 is a function of the velocity head and the pressure head intensity. Based upon the illustrations developed to this point, it should be noted that the head pressure has been created by gravitational force due to the weight of water contained in the gravity tank. The term total head pressure is used synonymously with total pressure intensity.

Different types of pumps or pressure tanks can be used to create pressure head and velocity head in contrast to the influence of gravity.

Pressure Designations

The examples presented in **Figures 2-15** and **2-16** represent several concepts associated with pressure measurement. But it should not be taken out of context for practical field measures of water supply capability.

Figure 2-17

Flow Tube Extension, Underwriter's Playpipe and Gauges

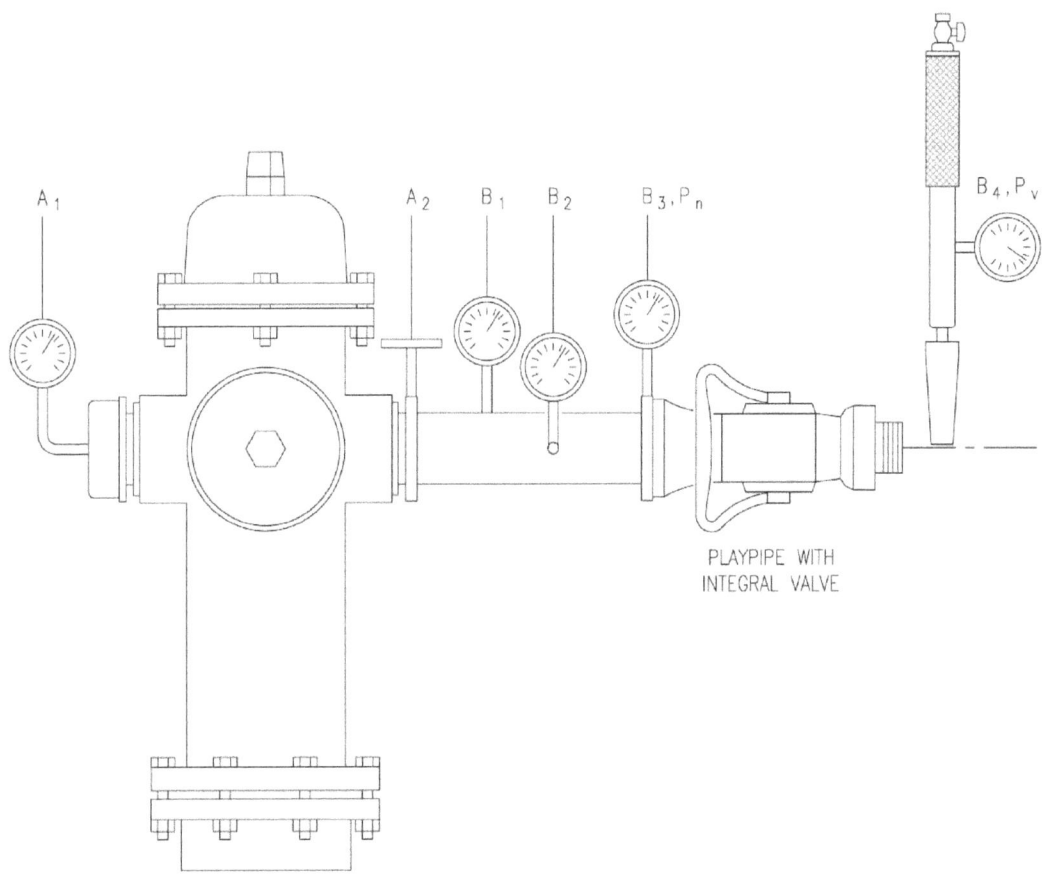

Examine **Figure 2-17**. Hydrant 1 in a small water supply distribution system (from **Figure 2-14**) is illustrated with equipment attached to the two 2-1/2-inch outlets. The equipment and devices are described below.

- A Bourdon gauge fire hydrant gauge A_1 attached to the left side of the fire hydrant outlet to read pressure directly in psi.

- A fire hydrant gate valve B_1 attached to the right side of the fire hydrant to control the flow to an Underwriter's playpipe and nozzle

⬥   The 25-inch playpipe is attached to the controlling gate valve. The tube length is equal to 10 times the diameter of the outlet orifice so that the flow turbulence will be minimal at the measuring point of the entrance to the nozzle tip. A Bourdon gauge is attached using an O-ring collector at location B_1 and location B_2.

⬥   At point B_3, a Pitot tube with a Bourdon gauge is attached to the playpipe by a holder so the orifice opening is located at a distance of 1.5 times the diameter of the nozzle tip orifice to a place where the water discharge stream will have the greatest mass continuity for accurate flow measurement.

⬥   Except in experimental hydraulic work, it can be assumed that the head loss or pressure loss across the nozzle tip is negligible for water supply testing purposes.

The Pitot tube, named for Henri Pitot who invented it in 1730 to measure velocity streamlines in pipe flow, is used currently to measure the pressure intensity of water discharging from the nozzle attached to the playpipe. The Pitot tube in **Figure 2-17** is correctly mounted using a clamping device to steady it. However, in flow testing work from fire hydrants on a municipal water system, hand-held Pitot tubes are used. It is very important that the orifice of the Pitot tube be placed in the discharge steam 1.5 times the diameter of the nozzle opening orifice diameter, or, in some cases the diameter of the fire hydrant outlet. Note: Refer to the text topic on Water Supply Testing Using Public and Private Water Supply Hydrants.

A fundamental understanding of what pressure function is being measured at each reference point in **Figure 2-17**, and how each of the reference pressures are related to flow measurement, is essential for the application of hydraulic theory to the testing and evaluation of municipal water supply systems.

NORMAL PRESSURE

Normal pressure is the pressure created by forces acting perpendicular to the pipe wall where a pressure tap is made. (Reference #31, pg. 47) This is actually an application of Pascal's concept that fluid pressure is perpendicular to the surface on which it acts. A simple pressure tap on a pipe or a cap gauge attached to a fire hydrant when the fire hydrant is opened but no water is flowing are perfect examples of this concept. In **Figure 2-17**, pressure taps are made regarding pressure at the cap gauge location A_1 and on the ring gauge collected to provide pressure to the gauge at location B_2. In Bernoulli's Formula, **normal pressure** represents P/w when the water is the liquid medium. It is necessary to understand that normal pressure is independent of the velocity flow in the fire hydrant, the underground pipe, or at the collector ring B_2 when water is flowing out of the open nozzle tip in **Figure 2-17**.

Usually, pressure is read from a recording gauge by means of simple treated adapter to the wall of a pipe. Pressure can be so read on cap gauges attached to fire hydrant as noted at reference A_1 in **Figure 2-17**, and where ring gauges are attached to the base of fire stream nozzles at B_3 in the same Figure. Normal pressure or the term **net pressure** as used in practical water system hydraulics becomes an important consideration in the design of water systems to know what the effect **velocity pressure** will have on individual fire hydrants and where taps are made on the water system piping to supply lateral lines to commercial and industrial properties; the impact of **velocity pressure** to dwellings typically is not of concern because the differential between normal pressure and velocity pressure is small due to the very low flow rate to dwellings at any specific time.

VELOCITY PRESSURE

Velocity pressure is a measure of the energy required to keep the water in motion in a pipe. (Reference #30, pg. 47) **Figure 2-17**, reference point B_4, illustrates the original concept of the Pitot tube as a device adapted to determining the local velocity of flow by measuring the pressure of the fluid at a point forward of the Pitot tube orifice inlet. The exact location is known as the vena contra location or the plane through the water where there is the greatest mass continuity leaving the nozzle. When the tube is placed at this location, which is one-half (2) the distance of the diameter of the nozzle or open hydrant butt on a fire hydrant, the pressure energy flowing up the tube and acting on the pressure gauge to give a reading in psi represents the total energy of the water in motion. **Figure 2-17** indicates that the velocity pressure head extends above the net pressure head when comparing the water level from the pressure tap and the water level from the Pitot tube inserted into the flowing stream.

Velocity pressure represents $V^2/2g$ in Bernoulli's equation. Since this equation uses unit of measure in feet, velocity pressure may be calculated by the formula:

Formula 19   Velocity head $(H)=V^2/2g$

To determine pressure in psi for water, the following conversions are used: (Reference #31, pg. 47)

Velocity pressure$=0.00673\,V^2$, derived as follows:

Velocity head, $H=V^7/2g$

or,

$psi=0.433\,V^2/64.32=0.00673\,V^2$

Where:

H=head in feet

psi=pounds per square inch

V=velocity in feet per second

g=acceleration due to gravity or 32.2 feet per second

For practical hydraulics associated with municipal water system a final formula can be $psi=0.00673xV^2$

However, a Pitot tube positioned in the center of a pipe with the open end of the tube points upstream against the flow will measure total pressure on the attached gauge. The symbol P(t) will be used to denote total pressure for calculations and reference. The difference between the total pressure P(t) read on the

Pitot gauge, and normal or net pressure P(n) on the gauge at nearly the same reference plane is, therefore, defined as velocity pressure. The proper relationship between total pressure, normal pressure, and velocity pressure is expressed by the following equations:

Total Pressure=Normal Pressure+Velocity Pressure

$$P(t)=P(n)+P(v)$$

Normal Pressure=Total Pressure-Velocity Pressure

$$P(n)=P(t)-P(v)$$

Velocity Pressure=Total Pressure-Normal Pressure

$$P(v)=P(t)-P(n)$$

PITOT PRESSURE

The term Pitot pressure has a different connotation in applied water supply hydraulics from that used in **Figure 2-17**. In applied hydraulic work, the term normally refers to the tube being used to measure pressure in a stream of water at the point of its discharge from the orifice of a hydrant outlet or a straight stream hose nozzle. This is illustrated at location B_4 in **Figure 2-17**. Note that the point of flow measurement is in open air, where the perpendicular pressure function no longer is a factor in the flow measurement. The Pitot tube in this application is held in the center of the circular stream of water. It is important to note that the rate of flow through the nozzle orifice is dependent on the size and shape of the orifice as well as the flowing pressure, the flowing pressure being defined as velocity pressure. In summary, the pressure read on the Pitot tube gauge is simply a function of velocity head. The mathematical relationship must be examined carefully. Improper use of the tube in flow measurement work can lead to inaccuracies in computations.

The use of the Pitot tube is based upon Torricelli's velocity equation: $V=(2gh)^{1/2}$. In water supply hydraulics, the term normally refers to the tube being used to measure pressure in a stream of water at the point of its discharge from the orifice of a hose nozzle, or a hydrant outlet (butt). This is illustrated at B_4 in **Figure 2-17**.

The device is also called a Pitot tube, but the point of measurement is in open air, where the perpendicular pressure function no longer is a factor in the flow measurement. The Pitot tube in this application is held in the center of the circular stream of water. It is important to note that the volume flowing through the orifice is dependent on the size and shape of the orifice as well as the flowing pressure; the flowing pressure being defined as **velocity pressure**. In other words, the pressure read on the Pitot tube is simply a function of velocity head. The mathematical relationship must be carefully examined. Improper use of the tube in flow measurement work can lead to inaccuracies in computation.

Problem 6

The stream velocity from a smooth-bore playpipe is calculated to be 100 feet per second (fps). What velocity pressure should be indicated on a Bourdon gauge attached to a Pitot tube held in the flowing stream?

Solution: Start with the component relationship on Bernoulli=s equation:

$$V=(2gh)^{1/2} \qquad \text{(Remember, h=2.31 p)}$$

Therefore: $V=[2(32.2 \times 2.31p(v))]^{1/2}$

$$V=(64.4 \times 2.31p(v))^{1/2}$$

$$V=(148.8)^{1/2}$$

Squaring both sides:

$$V^2=148.8$$

$$P(v)=10,000/148.8$$

$$P(v)=67.2 \text{ psi}$$

For the purpose of understanding the relationship between orifice pressure, as measured with the Pitot in **Figure 2-17** at location B_4 and the pressure measured at the base of the playpipe B_3, it should be noted that velocity pressure $P(v)$, and net pressure $P(n)$ become equal only under certain conditions. This means that $P(n)$ measured at location B_1, and $P(v)$ measured at B_4, can only be considered equal over a narrow range of flow conditions.

The important point to remember is that velocity pressure and normal pressure are two different measures of pressure intensity.

Additional pressure functions can be identified with **Figure 2-17**. These other pressure functions in a water system can be illustrated as follows:

Set 1: Hydrant discharge valve A_2 is closed, so no water will flow from the playpipe. The hydrant is opened so that the static pressure registers on the hydrant cap gauge. Based on previous information given with **Figure 2-16**, the static pressure reads 65 psi. Remember, static pressure is pressure from the weight of water at rest; that is, no flow. In this example, the static pressure is determined from the height of still water in the gravity tank.

Set 2: Now assume that hydrant discharge valve A_2 is opened so that water discharges through the flow tube and the playpipe with a discharge of 250 gpm flowing from the nozzle. Each of the gauge readings at A_1, B_1, B_2, B_3, and B_4 will reflect a pressure function based on the terms defined: total pressure, velocity pressure, or normal pressure. However, each of the pressure functions has another significance in terms of flow.

The water is flowing and a hydrostatic condition has changed to a hydrokinetic condition. This fact leads to some new, important reflections relative to pressure intensity at each of the defined reference points. The following numerical values are used for illustrative purposes only, and may not reflect the true hydraulic condition of a specific water supply system.

Using the assumptions in Set 2 above, the following pressures are read at each of the reference points. Remember that the assumed flow condition is 250 gpm from the nozzle.

Location A   (1): Static Reading = 65 psi

When 250 gpm are flowing from the nozzle, the pressure reading on the cap gauge drops to 55 psi. This new reading is defined as **residual pressure**.

Residual Pressure:

Residual pressure or dynamic pressure is the pressure at a given point in a pipe or fire hydrant with a specific volume of water (i.e., 250 gpm) flowing through the pipe. (Reference #32, pg. 47)

To be meaningful, when the residual pressure is stated, the volume in gpm or million gallons per day (gpd) must be stated with the reference static pressure, as well as the horizontal and vertical position where the pressure is measured.

Pressure Drop:

Pressure drop or pressure loss is the difference between static and residual pressure at the same point, A_1. In the example under consideration, a static pressure of 65 psi dropped to a residual pressure of 55 psi, when 250 gpm are measured at the discharge from the nozzle. In this case, the 10 psi pressure drop is a measure of the friction loss due to the established flow of water in the system of supply piping to the point where the gauge is read.

Pressure Differential:

Pressure differential, usually denoted as $\triangle P$, signifies the difference in residual pressures for any two known gpm flows.

Using **Figure 2-17** for an example, the size of the orifice on the playpipe might be increased from a 1-1/8-inch tip to a 1-1/4-inch tip. The corresponding flow may increase from 250 gpm to 300 gpm. This increased flow will lower residual pressure at A_1 to an assumed 45 psi. The pressure differential, $\triangle P = 55$ psi-45 psi, or 10 psi. This concept becomes useful and important in calculating potential flow from hydrants during fire flow tests.

Pressure Profiles

The examples used in the text material on Hydraulic Fundamentals Applied to Water Supply Systems use a gravity tank as the source of supply for the water in a single supply main to feed the fire hydrants and a standpipe system in a factory building. This is typical of the water supply components associated with a small water distribution system. However, a municipal water supply system is required to supply daily needs for domestic consumption and commercial consumption, **plus** meet required fire flows for a working fire situation.

The gravity tank in the illustrations used is a true static supply source, because the water in the tank is at rest; no water is being drawn from the system until a hydrant is opened, or a hoseline attached to the standpipe in the illustrated building is opened. However, water in a municipal water system presumably is flowing at all times to supply domestic needs. Therefore, what is called city main static pressure is only a **reference pressure** that is affected by the supply for consumers at any given time. At any point of measurement, for example at a fire hydrant, the pressure reading on a Bourdon-type cap gauge reflects the hydraulic profile pressure at that point, based on the simultaneous water flow in the underground pipe for domestic services. A common practice in water distribution system analysis is to record profile pressures at a number of locations on the distribution network to determine the percentage of low contribution by primary and secondary feeder lines.

It is important to note that, unless water tests are being conducted on a closed system (i.e., one that is not supplying domestic or industrial consumption), the cap gauge reading on a given hydrant represents a profile pressure. The profile pressure is only meaningful for determining available water supply for fire protection purposes when evaluated in accordance with corresponding domestic consumption.

References:

1. Henke, Russell W. *Introduction to Fluid Mechanics*. Reading: Addison-Besley Publishing Co., 1960, pg. 2.

2. ibid., pg. 3.

3. Wood, Clyde M., CE. *Automatic Sprinkler Hydraulic Data*. Cleveland: Automatic Sprinkler Corporation of America, pg. 3-1.

4. ibid.

5. ibid.

6. Henke, *op. cit.*, pg. 3.

7. Henke, *op. cit.*, pg. 4.

8. ibid.

9. Wood, *op. cit.*, pg. 3-1.

10. Russell, George E. *Textbook on Hydraulics*. New York: Henry Holt and Company, 1934, pg. 65.

11. Henke, *op. cit.*, pg. 4.

12. Henke, *op. cit.*, pg. 5.

13. Russell, *op. cit.*, pg. 8.

14. Rusk, Roger D. *Introduction to College Physics*. New York: Appleton-Century-Crofts, Inc., 1974, pg. 2.

15. ibid.

16. Henke, *op. cit.*, pg. 6.

17. Henke, *op. cit.*, pg. 7.

18. Bardsley, Clarence E. *Historical Resume of the Development of the Science of Hydraulics*. Stillwater: Publication No. 39, Vol. 9, No. 6, April 1939.

19. Rusk, *op. cit.*, pg. 5.

20. ibid.

21. ibid.

22. Henke, *op. cit.*, pg. 47.

23. ibid.

24. ibid., pg. 48.

25. ibid., pg. 567.

26. Rusk, *op. cit.*, pg. 168.

27. Wood, *op. cit.*, pg. 3-4.

28. King, Horace, and Ernest Brater. *Handbook on Hydraulics for the Solution of Hydrostatic and Fluid Flow Problems.* New York: McGraw-Hill Book Company, Inc., pg. 6.

29. *ibid.*, pg. 4.

30. *ibid.*, pg 31.

31. Wood, *op. cit.*, pg. 3-11.

32. *ibid.*

Chapter 3: Evaluating Existing Water Supplies for Fire Protection Delivery Capability

A Water Supply Effectiveness Issue

Fire marshal's offices and insurance adjuster's offices are filled with reports that read like the following:

The LOSS REPORT: Fire was discovered on the second floor of the Hamilton Elevator Company at 10:45 p.m. The second floor served as the product finishing area and painting area before the elevator cage components were assembled. Fire was showing through an outside window on the street side when the first Engine Company arrived at the fire site. A 2-1/2 inch line was stretched to the fire floor for the initial fire attack. However, the lack of water from the fire hydrant supply resulted in a weak and ineffective fire steam. Due to the lack of an effective water supply the interior of the building was a total loss. The loss to the building is estimated to be $960,000.00 and the estimated content loss was set at $780.000.00 by the Insurance Adjuster. (American Mutual Insurance Alliance) (Note: The company name has been changed to prevent specific identification in accordance with privacy legislation.)

Without adequate and reliable water supplies at fire hydrants to protect fire risks, the best-trained firefighters with the best of equipment have only a very limited opportunity of protecting lives in a building fire and confining, controlling to the area of origin, and extinguishing a hostile fire. Therefore, it is essential that community officials, from the highest ranking administrator/manager, to the water department superintendent or an equal authority, and the ranking fire chief, provide means and opportunity to monitor the community water system's performance capability constantly. (Reference #1, pg. 65)

Determining Existing Community Water Supply Adequacy and Reliability

A community water supply system is one of the most important factors in both public and private fire protection. Fire departments and fire protection engineers, as well as those responsible for the design, operation, and maintenance of water systems, are concerned with two aspects of the total water supply system: its adequacy and its reliability.

Adequacy, in the case of a water system supplying water for normal consumer consumption and for fire protection, means having the capability of simultaneously supplying water for maximum consumption demands plus water that may be needed to combat and extinguish a major fire within the area served by the water system. Adequacy concerns itself with sufficient flow and pressure on all installed fire hydrants on the water system; the minimum residual pressure on each fire hydrant under flow conditions is to be 20 pounds per square inch (psi) residual pressure. (References #2 and 3, pg. 65)

Reliability of a community water system is having the capability of supplying the maximum daily consumption plus a required fire-flow demand, even in the event of a malfunction or the outage of important system components, such as a pipeline break, valve failure, power outage, or stationary pump outage. Reliability is a more subjective evaluation and requires both a **what-if** look at the water system and a determination of what to do about the **what-if** happening. (Reference #4, pg. 65)

Today, the reliability of community water systems has to be extended to the consideration that the water supply sources maybe contaminated through terrorist operations or depleted through overt operations.

This topic examines the objectives of water supply testing, using fire hydrants to determine water supply capability throughout a given water distribution system, some applications of fundamental hydraulics (introduced in Chapter 2), flow test procedures, and graphical solutions to test for water flow problems. The results of this type of analysis are essential to understanding a given water system's capability to provide both consumer consumption and needed fires flows at representative locations throughout the built areas of the community.

FLOW TEST OBJECTIVES

Fire hydrant flow tests conducted on public water supply systems are done for several reasons: 1) to determine the rate at which water is available at specific locations within a given distribution system; 2) to use flow-test data between two fire hydrants on the same water main to determine a pipe coefficient or to determine if control valves are completely open; 3) to determine water availability at the end of an existing pipeline for the determination of pipeline extensions; 4) for determining the need for booster pump applications; 5) to verify or calibrate the accuracy of water distribution system models; and 6) to determine a water flow and pressure profile where the water system supplies an automatic sprinkler system. The flow-test data may be used for the evaluation of new developments that would be supplied by the water system and for evaluating tradeoffs for providing water supplies for public fire protection and/or private fire protection in the form of automatic sprinkler systems.

Of particular interest to fire departments and insurance companies is the rate and quantity at which water is available to concentrated high-value areas, such as shopping centers, industrial parks, highrise high-tech buildings, institutional buildings, and residential areas. (Reference #5, pg. 65) Also see National Fire Protection Association (NFPA) 25, *Standard for Inspection, Testing, and Maintenance of Water-Based Fire Protection Systems*, which outlines testing requirements for private fire hydrants. (Reference #5, pg. 65)

Fire hydrant flow tests should not be attempted until all the operational characteristics of a water system are known. Results may differ substantially, depending on the operation of pumping equipment, water levels in the system's storage facilities, rates of consumption, and points of demand on the water system. Even though it is possible to conduct accurate tests within acceptable tolerances, often the results obtained will vary from day to day, and even at different periods during the same day, because of the many variables involved.

BASIC HYDRAULIC CONCEPTS

Persons involved in water supply testing and flow testing from fire hydrants need to understand some of the fundamentals of water as a fluid flowing through pipe and flowing out of orifices such as an outlet on a fire hydrant. The Hazen-Williams Formula for evaluating water flow though pipes is the most practical and usable formula for analyzing water supply systems in relation to providing effective water supplies for fire protection. (Reference #6, pg. 65) Many standards published by the NFPA, including those on sprinklers, water spray systems, and suburban and rural water supplies, make direct reference to the Hazen-Williams Formula for pipe configuration calculations. It is the formula of choice for water system operators and field engineers to measure pressure loss in pipe sections and to verify the "c" value or the coefficient of roughness on the interior of pipe walls which signals a reduction in the actual pipe diameter. These concepts will be discussed further below. (Reference #7, pg. 65)

The **Hazen-Williams Formula** is used widely for pipe flow problems involving municipal water supply system evaluations and sprinkler system piping design layout and evaluation. This is an empirical formula that evolved for water test work over a period of 30 years and is considered to be valid today for water distribution system analysis. The formula is presented as follows: (Reference #8, pg. 65)

$$P = 4.52 \times Q^{1.85} / c^{1.85} \times d^{4.87}$$

Where:

P=pressure loss in psi per foot of pipe, often refereed to as friction loss.

Q=flow of water in U.S. gallons of water per minute expressed as gpm.

c=roughness coefficient to be used with this formula; see further explanation below.

d=the actual internal diameter of the pipe; for practical hydraulics the published diameter of the pipe is used, not the actual manufacturers' diameter. (A given brand of 6-inch pipe has an actual internal diameter of 5.871, which is indistinguishable from field hydraulic problems.)

NOTE: on a special understanding the correct use of this formula: If the constant for the pipe which is now an acceptable constant for all pipe from the perspective of practical hydraulics is canceled out, and the coefficient of roughness (c) is canceled out, and the diameter of the pipe being tested remains constant (i.e., all 8-inch pipe), it too can be canceled out. The remaining equation now reads

psi loss varies as $Q^{1.85}$

This concept is very important because it permit preparing Log graphs that are based on this equation. This is useful for plotting flow data to show the relationship between flow and pressure loss through pipelines without doing a lot of mathematical calculations. **Figure 3-1** illustrates this type of graph labeled as a Water Flow Test Summary Sheet. This type of sheet will be used in sample problems.

Figure 3-1

Water Flow Test Summary Sheet

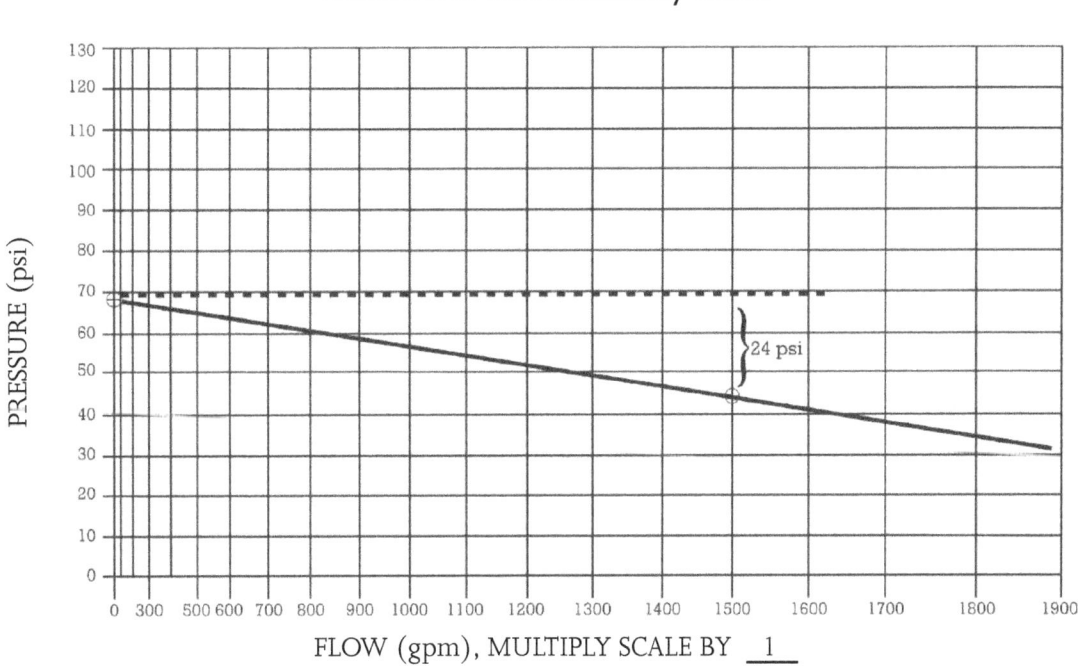

The calculation of friction loss in a pipe section depends on the quantity of flow in gpm, the roughness coefficient and the internal diameter of the pipe. Several methods can be used for solving the formula: 1) straight mathematical computation; 2) the use of power tables; or 3) the use of nomograph (e.g., hydraulic slide rules). With the use of scientific calculators and computer hydraulic calculation software calculations are easier.

The basic Hazen-Williams Formula above can be transposed as given below to solve for the "c" pipe condition or roughness factor. Actually, friction loss in a pipe section cannot be determined until the pipe coefficient is known or estimated. To improve the accuracy of computations, it is often desirable to conduct flow studies to determine the actual "c" value for a given pipe section, a representative "c" value for the pipe system. The friction loss is determined from a field study with a know quantity of water flowing in a given diameter of pipe. The empirical field value then can be used to solve for "c" as follows:

$$c = 2.26Q/(d^{2.63}xp^{0.54})$$

A number of laboratory and field studies were conducted in the in the 1970s by Factory Mutual (FM) Systems to determine representative "c" values for different pipe conditions flowing water. (Reference #9, pg. 65) **Table 3-1** depicts a number of "c" values that can be used for hydraulic calculations or to conduct comparative "c" values between existing conditions and the table values.

Table 3-1

Hazen-Williams Pipe Coefficients*

Kind of Pipe	Value of C		
	1	2	3
Cast iron, unlined, public mains or fire-protection mains with mill-use draft			
10 years old	105	90	75
15 years old	100	75	60
20 years old	95	65	55
30 years old	87	55	45
50 years old	75	50	40
Cast iron, unlined, new	120		
Cast iron, cement-lined	130		
Cast iron, bitumastic-enamel lined	140		
Cement-asbestos	140		
Average new steel	140		
Sprinkler piping, black or galvanized:			
Over 2 in	120		
2 in or less	100		
1. Water mildly corrosive. Use same values for fire-protection mains having no mill-use or domestic draft.			
2. Water moderately corrosive.			
3. Water severely corrosive, including well water.			

*Factory Mutual Systems, *Handbook of Industrial Loss Prevention*, New York: McGraw Hill Book Company, 1967, p. 21-9.

MEASURING WATER FLOW FROM SMOOTH-BORE ORIFICES

The original experimentation on flow from smooth-bore orifices was conducted by the National Board of Fire Underwriters in 1912. (Reference #10, pg. 65) This work led to the creation of the standard playpipe, or the Underwriters playpipe, as illustrated in **Figure 3-2**.

Figure 3-2

Underwriters Playpipe

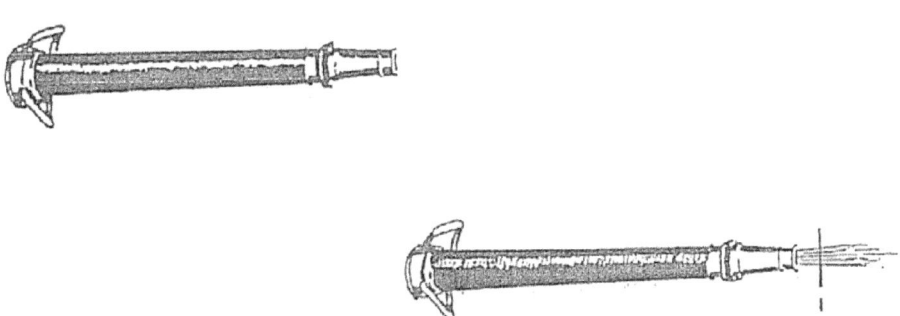

Playpipe nozzle and measurement of orifice pressure

John R. Freeman adapted a Pitot tube device to be used in measuring flow from the Underwriters playpipe. The follow formula is known both in hydraulic textbooks and in fire service publications as the Freeman Flow Formula or the Underwriters Flow Formula for measuring flow in gpm from smooth-bore fire-service nozzles and fire hydrant outlets. The final version of the discharge formula is as follows: (Reference #10, pg. 65)

gpm=29.83 Cdxd^2(Po)^2

Where:

Cd=a coefficient for the nozzle or fire hydrant outlet type being used.

d=the measured diameter of the orifice where the water is discharged.

Po=orifice pressure measured by a Pitot tube placed in the flowing steam at a distance one-half the diameter of the orifice; technically called the vena contracta point to achieve the most accurate flow reading.

Figure 3-3 provides illustrations of typical Pitot tubes and the proper position of the Pitot tube in the waterway; this is considered to be one-half the distance of the orifice opening.

Figure 3-3

Illustrations of Typical Pitot Tubes

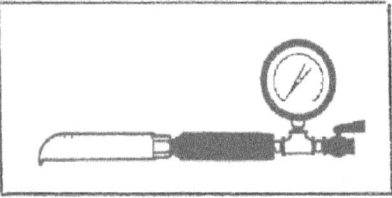
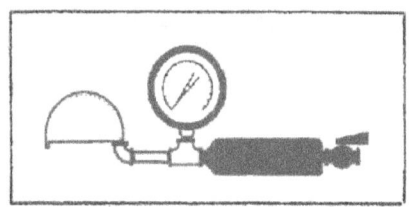

Air Chamber Type   **Pressure Snubber Type**

Therefore for the 2-1/2-inch standard opening on a fire hydrant, the opening on the blade to the Pitot tube should be approximately 1-1/4 inches from the hydrant opening.

Also of importance is the coefficient of discharge from different hydrant outlets. **Figure 3-4** illustrates how the discharge nipple on hydrants may be installed. In hydrants produced over the past 50 years, three types of nipple insertions have been used by different hydrant manufacturers. Each has a different coefficient of discharge that needs to be used in the Underwriters Flow Formula (also known as the Freeman Flow Formula). This is illustrated in the following example of computing the available flow from a hydrant outlet.

Figure 3-4

Typical Fire Hydrant Outlet Coefficients

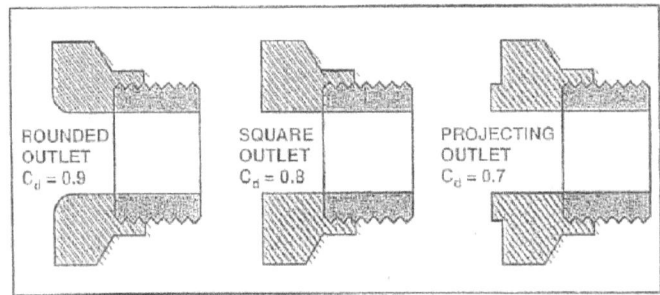

Example 1: Determine the discharge flow in gpm from a hydrant with a flowing 2-1/2-inch outlet that has a coefficient of discharge that is 0.90 and the flowing Pilot pressure is $16^{1/2}$ psi.

- Step 1: The formula is $gpm = 29.83 \times Cd \times d^2 \times (Po)^{1/2}$

- Step 2: Insert the given values for *Example* 1 in the formula:
  $gpm = 29.83 \times 0.90 \times (2.5)^2 \times 16^{1/2}$

- Step 3: $gpm = 29.83 \times 0.90 \times 6.26 \times 4$

⬥ Step 4: Or, 29.83x0.90x25.04

⬥ Step 5: Answer: 672.25, or 672 gpm

The American Water Works Association (AWWA) has developed tables that can be used to read flows directly based on the above formula and all of the variables.

The Underwriters Flow Formula may be used to calculate the discharge from other fire protection equipment and pipe by using a typical discharge coefficient for solid-stream nozzles and devices. The **key words** here are **solid stream**. It is essential that the entire perimeter of the discharge orifice be **wetted**. In other words, water must be flowing full through the orifice, which requires a good stream of water. Partial flows can not be calculated accurately using the above formula because there is a cavitation in the stream diameter flow. A full-flowing outlet is essential when the hydrant is initially opened; then close the fire hydrant and attach a smooth-bore nozzle. Try tip sizes until the pressure reading is between 30 and 80 psi; the ideal pressure range is 40 to 60 psi.

Table 3-5 presents discharge coefficients for different types of outlets.

Table 3-5

Typical Discharge Coefficients of Solid-Stream Nozzles*

Outlet	Discharge Coefficient
Standard sprinkler, average (nominal 1/2-inch diameter)	0.75
Standard orifice (sharp edge)	0.62
Smooth-bore nozzles, general	0.96 – 0.98
Underwriters playpipes or equal	0.97
Deluge or monitor nozzles	0.997
Open pipe, smooth, well-rounded	0.90
Open pipe, burred opening	0.80
Hydrant butt, smooth, on well-rounded outlet, flowing full	0.90
Hydrant butt, square and sharp at hydrant barrel	0.80
Hydrant butt, outlet square, projecting into barrel	0.70

*U.S. Army Engineering School. *Hydraulics II*. Fort Belvoir: Author, Sub-course No. 433, March 1969, p. 3-5.

CONDUCTING FIRE-FLOW TESTS

1. General considerations.

   Fire-flow tests are conducted to determine pressure and flow-producing capabilities at any location within the distribution system. The primary function of fire-flow tests is to determine how much water is available for fighting fires, but the tests also serve as a means of determining the general condition of the distribution system piping network. For example, heavily tuberculated water mains, or those with heavy wall deposits can reduce the flow-carrying capacities of pipe; this reduced capacity can be detected by means of flow tests. Flow tests also can help detect closed valves in the system. The results of flow tests are used extensively by insurance underwriters as a factor in setting property insurance rate premiums; they also are used by designers of automatic sprinkler systems. (Reference #11, pg. 65)

   It is good practice to conduct flow tests on all parts of the distribution system annually and at representative fire risks semiannually to identify the service areas that might be affected by any significant change in the distribution system piping and accessories such as control valves. All problem conditions should be investigated immediately and the proper corrections made in a timely manner.

   An accurate flow graph as discussed below should be kept on file at the water department office for each flow test conducted. Each graph line should be dated. A copy of each flow graph with the updates should be provided to the fire chief for distribution to the first-responding companies in specific areas of the community.

2. Operational definitions of terms used in water flow testing. (Reference #12, pg. 65)

   a. **Flow Hydrant**: The hydrant or hydrants where flow is actually measured.

   b. **Pitot Pressure**: The pressure reading obtained on the Pitot gauge during a flow test.

   c. **Pitot Tube**: An instrument, as discussed above, that is used to measure the flow of water discharged from a fire hydrant outlet (orifice) by measuring and converting velocity head into a pressure head reading on a gauge in psi. See **Figure 3-6**.

   d. **Residual Pressure**: The pressure in the distribution system piping, measured at the residual pressure fire hydrant, at the time the flow readings are taken at the flow hydrant(s).

   e. **Static Pressure**: The pressure that exists at a given point under normal distribution-system flow conditions.

Figure 3-6

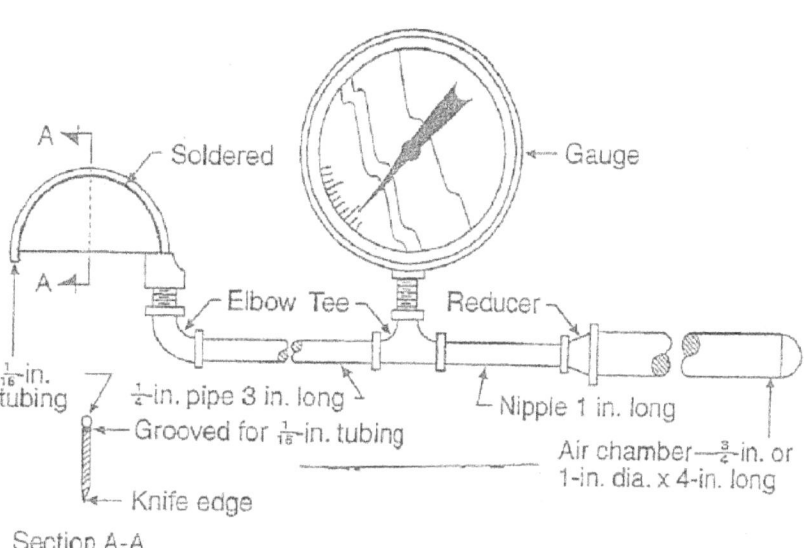

3. Personnel and equipment for flow tests. (Reference #13, pg. 65)

    The following list includes the required personnel and equipment needed to conduct a flow test. Equipment should be in good working order and be available at the time of the test.

    a. For each designated flow hydrant, one Pitot tube with a calibrated pressure gauge capable of reading 0 to 60 psi.

    b. One outlet-nozzle cap that will fit the outlet nozzle of the residual fire hydrant; often referred to as a cap gauge. (Refer to **Figures 3-6** and **3-7**.) The outlet nozzle is equipped with a pressure gauge capable of reading from 0 psi to 25 psi greater than the pressure expected at the residual fire hydrant.

    c. A ruler, graduated to at least 1/16 of an inch, to measure the inside diameter of the outlet nozzle of each flow hydrant.

    d. One hydrant wrench to operate the residual fire hydrant and one to operate each of the fire hydrants at which the flow will be measured.

    e. One discharge diffuser to absorb the energy from the hydrant flow so that it is contained, where necessary, to avoid property damage or to minimize the effect on traffic.

    f. One person to read the gauge on the residual hydrant and one person to read the gauge on the Pitot tube for each of the flow hydrants.

g. Clipboards and sheets for recording data at each fire hydrant.

h. For wet-barrel hydrants found in the Sun Belt regions, it may be necessary to install a specially designed nozzle to minimize turbulence caused by the discharge valve.

Special Note: The Pitot tube and the pressure gauges are relatively delicate instruments and must be treated accordingly. Gauges should be checked for accuracy at reasonable intervals to ensure that the flow tests will be as accurate as possible.

Figure 3-7

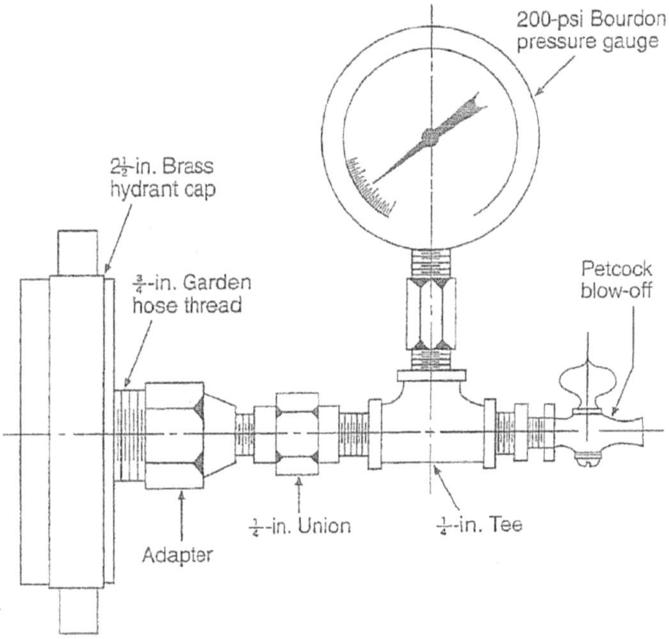

4. Preplanning prior to conducting field tests.

a. Review **up-to-date** water system distribution maps and determine which hydrants will be used to measure flow and which will be used to measure the static and residual pressures according to the **suggested flow-test locations** depicted in **Figure 3-8**. All fire hydrants should be at approximately the same elevation. Otherwise test results may need to be corrected for elevation differences.

Figure 3-8

Suggested Flow-Test Locations

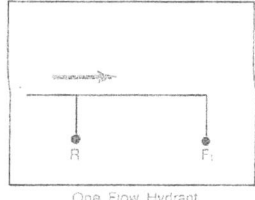

One Flow Hydrant

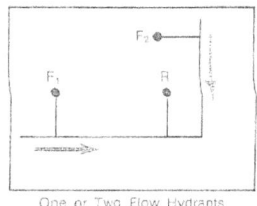

One or Two Flow Hydrants

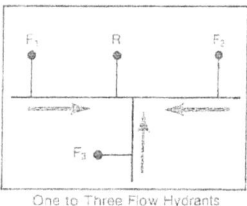

One to Three Flow Hydrants

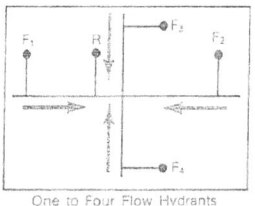

One to Four Flow Hydrants

Note that the arrows indicate direction of flow: R–residual fire hydrant; F–flowing fire hydrant.

b. Review previous tests to estimate the flow and pressures that can be expected.

c. Select a day for testing when system consumer consumption will be normal and weather predictions indicate that conditions will be reasonable. The operating division should be notified as to the time and location of the tests so necessary adjustments to the system can be made. Investigate traffic patterns, as the tests may affect traffic flow. It may be necessary to notify the traffic division of the police department to reroute traffic for the short period that fire-flow tests are being conducted.

5. Field procedures for flow tests.

a. Make provisions for minimizing interruption to traffic to the extent possible and for adequate discharged drainage of water to avoid property damage.

- Locate the residual fire hydrant and do the following:
    -- Flush the residual fire hydrant to eliminate sediment that may damage the gauge.
    -- Install the outlet-nozzle cap equipped with the pressure gauge on a hydrant nozzle; preferably on the side where the reader can observe the flowing fire hydrant.
    -- Open the main valve **slowly** until all the air is vented. Close the vent and open the main valve fully.

-- Read the gauge; when the needle has stabilized, record the reference static pressure.

- Locate the flow hydrant(s) and do the following:

-- Measure and record the inside diameter (ID) of the outlet nozzle from which the flow is to be measured.

-- Determine the outlet nozzle coefficient in accordance with **Figure 3-4**. The coefficient allows for differences in hydraulic entrance losses. The first illustration is the most common assembly and represents a rounded shoulder at the entrance. If the configuration of the hydrant to be tested differs significantly from the configuration shown in **Figure 3-4**, contact the hydrant manufacturer for the coefficient.

b. Conduct the flow test as follows:

- Station one observer at the residual hydrant and one observer at each flow hydrant.

- Open each flow hydrant slowly until it is fully open. Open one fire hydrant at a time to avoid a pressure surge.

- When the pressure at the residual fire hydrant has stabilized, the observer signals the persons stationed at the flow hydrants to take the readings. The readings for residual pressure and the Pitot-tube readings for each flow hydrant must be taken simultaneously. The air should be exhausted from the flow hydrant before the reading is taken. For an accurate reading, hold the Pitot tube in the center of the nozzle, with the axis of the Pitot tube opening parallel to the direction of flow. The Pitot tube should be held away from the end of the nozzle at a distance of about half the nozzle diameter.

- Record the residual reading and the Pitot-gauge reading at each flow hydrant. Then close the flow hydrants one at a time.

For reasonably accurate test results, the pressure drop between the static and the residual pressures should be at least 10 psi; a 50-percent drop is considered to be the most accurate test, since at this drop on any given system the mean velocity of the underground main should be reached. Additional flow hydrants, if possible, should be added to achieve this condition.

It is best for observers to calculate the flow in the field so that, if the results appear in error, the test can be repeated immediately.

CAUTIONS TO BE OBSERVED WHEN FIELD TESTING

Opening a fire hydrant rapidly can cause a negative pressure fluctuation. Therefore, fire hydrants should be opened slowly until fully opened. Closing the hydrants is more critical, and it must be done very slowly until after the flow is diminished to about 20 percent of the full flow. Closing a hydrant rapidly causes a pressure surge, or water hammer; this could cause weakened water mains to **fail.**

Fire hydrants should be opened and closed one at a time to minimize the effect on the water distribution system. Dry-barrel hydrants must be opened fully because the drain valve mechanism operates with the main valve. A partially opened fire hydrant **could** force water thorough the drain outlets under pressure, eroding the thrust block support from behind the fire hydrant. After the test, the hydrant barrel should be drained before tightening the outlet-nozzle cap; a tight outlet-nozzle cap could prevent proper drainage and possibly cause ice blockage in either the upper or lower barrels.

Gauge measurements should be taken only when the water is running clear, because sediment could damage the instruments.

EXAMPLES OF AVAILABLE FIRE-FLOW DETERMINATIONS

Just maybe, the preparation of **Water Flow Test Summary Charts** using **Log N1.85** graph paper prepared on the basis of the Hazen-Williams Flow Formula to depict flow and pressure characteristics for installed fire hydrants and other fire protection devices, including automatic sprinkler systems, **is** the most efficient and effective tool for the evaluation of water supplies. There are several supply sources for this paper; one is referenced below. To establish a better appreciation of the value flow-test summary sheets can add to the evaluation of community water supply systems, three examples are given below:

1. Flow Analysis Problem #1 with associated flow test **Diagram 3-1**.

Diagram 3-1

Water Flow Test Summary Sheet

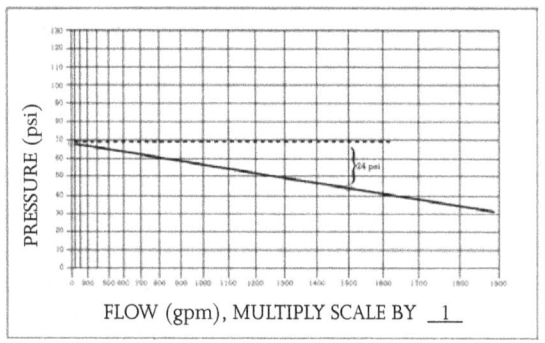

FLOW (gpm), MULTIPLY SCALE BY __1__

One Flow Hydrant

The diagram illustrates a typical flow-test arrangement for a dead-end water main. In other words, the water main flow does not extend beyond flow hydrant 1. The residual pressure is measured at hydrant R. The following provides the collected field information:

❧ Location R: static pressure = 65 psi.

❧ Residual pressure = 30 psi.

❧ Location F1: Flow from a 2-1/2-inch open butt on the fire hydrant indicates a flow of 498 or 500 gpm as measured by a Pitot and the flow determined from charts by or orifice calculation (see example).

❧ Using the left psi pressure line, the static pressure of 65 psi is recorded.

❧ Again, moving down the left psi pressure line find a residual pressure of 30 psi; move right along this line until the flow in the underground pipe of 500 gpm vertical point is reached; plot this point.

❧ Next, connect the dots between 65 psi static pressure and 500 gpm at 30 psi residual pressure and extend the line down to the baseline that has the marked flow values.

❧ Go to the 20 psi pressure line; indexed left. Read right until this line intercepts the diagonal flow line. Read a flow of 565 gpm at 20 psi residual pressure. Understand that this is the calculated potential flow from the **R** hydrant in **Diagram 3-1**. The flow hydrant downstream of the residual pressure/ flow hydrant is used to calculate the water system flow at this location to avoid turbulence in the water main for the best accuracy possible under field conditions. For comparative purposes in the future, the bottom line indicates that there is essentially no water system pressure when the flow is calculated to be 600 gpm.

2. Flow Analysis Problem #2 with Associated **Diagram 3-2**.

Diagram 3-2

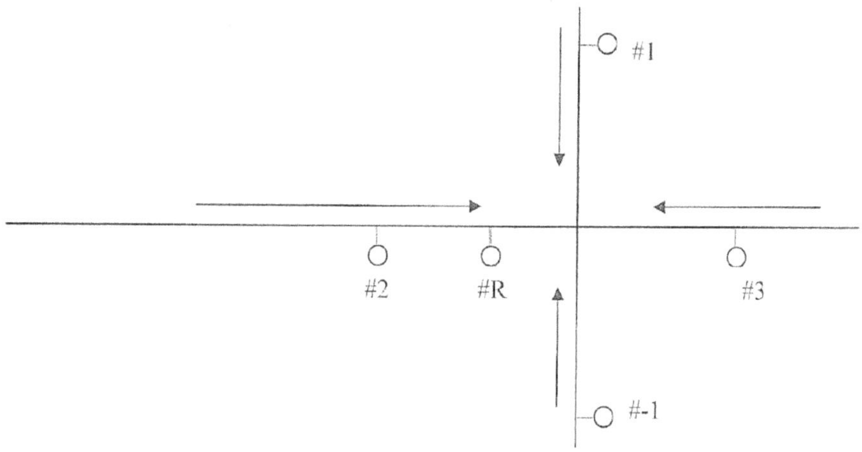

Four Flow Hydrants

In high-value commercial districts of a community, water may feed to an intersection to provide fire protection along the street fronts in each direction from the intersection. In this case it is useful to conduct what is called an area fire-flow test. This test will indicate the potential amount of water flow from four fire hydrants that provide water for four mobile pumps in case of a serious fire that has the potential for spreading between buildings.

Problem # 2 follows the same logical analysis as Problem #1 above. Therefore the following points summarize the calculations at 20 psi residual pressure:

⬧ Hydrant F1 = 650 gpm.

⬧ Hydrant F2 = 875 gpm.

⬧ Hydrant F3 = 950 gpm.

⬧ Hydrant F4 = 1,150 gpm.

⬧ The pressure is dropped on the water system around the intersection of pipe as illustrated in **Diagram 3-2** to **no less** than 20 psi residual pressure on any one flow fire hydrant. When the **residual** gauge reading becomes stable, flow readings are taken with a Pitot tube on all four flowing hydrants. These readings are converted into gpm flow. Finally, the 20 psi should be maintained on a fire hydrant that is supplying a mobile pump to avoid cavitation in the underground pipe network.

Problem #2 summary: The total available fire flow around the subject intersection is 3,625 gpm at 20 psi residual pressure.

Problem #3 is concerned with water supply to a structural property that is protected by an automatic sprinkler system. It is essential that sprinkler pressure at the alarm valve or dry pipe valve be maintained during a fire. There also may be the need to supplemental hose streams to complete extinguishment of the fire while the sprinklers are operating.

Figure 3-11

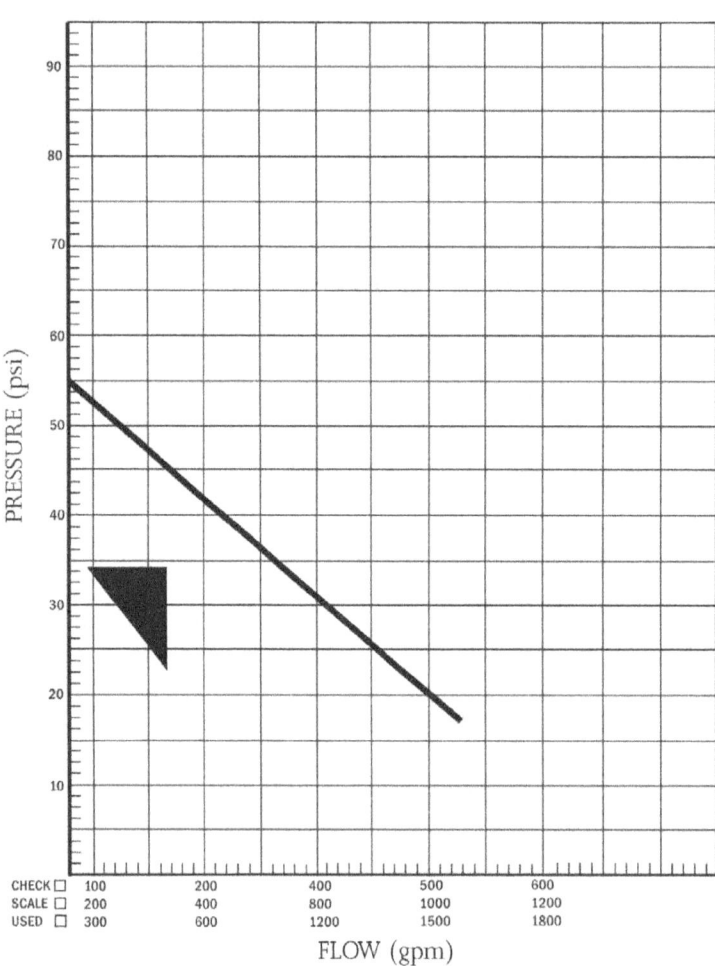

In this problem, a commercial fire risk at 140 Adams Street in Yourville has an installed sprinkler system. The calculated sprinkler system demand is 170 gpm at 34 psi at the street connection. The water system static pressure at this location is 55 psi. A flow test near this location indicates a flow of 500 gpm at 20 psi. This information is shown as **Figure 3-11**. The flow diagram depicts the amount of water and the sustained pressure for the sprinkler system in the shaded area to the left. Reading down the curve, it is determined that only approximately 100 gpm can be used for a supplemental hose stream. An excessive amount of water will cause a diminishing pressure on the sprinkler system.

SUMMARY

This topic is about conducting evaluations of municipal water systems by conducting water supply tests at regular intervals; semiannually is recommended. Most important is the concept of preparing a *Water Flow Test Summary Sheet* for each test location, along with the information discussed above. These sheets are very important for monitoring the municipal water system at specific locations over time. When flow curves at the same location over succeeding tests do not match, there is a need to know why they do not match and identified problems need immediate attention.

Responding fire companies to specific fire risks need to have current information about water supplies. Therefore, all first-due fire companies should have a set of flow test sheets to make informed decisions on fire suppression tactics. This type of evaluation should be part of any cost reduction program.

References:

1. Accident and Prevention Department. *Simplified Water Supply Testing for Fire Departments and Insurance Engineers*. 4th Ed., Chicago: American Mutual Insurance Alliance, 1970, pg. 4.

2. Cote, Arthur E., Ed., *Fire Protection Handbook*. 19th Ed., Quincy: NFPA, 2005, pg. 10-97.

3. Insurance Services Office. *Fire Suppression Rating Schedule*. Jersey City: Author, 2005, pg. 34.

4. *Fire Protection Handbook*, 19th Ed., pg. 10-97.

5. ibid., Chapter 10, Section 6, by Gerald R. Schultz.

6. Hickey, Harry, E. Ph.D. *Hydraulics For Fire Protection*. Boston: NFPA, 1980, pg. 82.

7. ibid., pg. 93.

8. ibid., pg. 95.

9. ibid., pg. 94.

10. Shepperd, Fred. *Fire Service Hydraulics*. Case-Shepperd-Mann Publishing Corporation, 1941, pg. 35.

11. American Water Works Association. *Manual of Water Supply Practices, Installation, Field Testing, and Maintenance of Fire Hydrants, AWWA–M-17*. 3rd Ed., Denver: Author.

12. ibid., pg. 39.

13. ibid., pg. 40.

14. The International Fire Service Training Association. *Water Supplies For Fire Protection*. 3rd Ed.--IFSTA 205. Stillwater: Fire Protection Publications.

CHAPTER 4: EVALUATING MUNICIPAL WATER SYSTEM STORAGE

FUNDAMENTAL CONSIDERATIONS

Municipal water supply systems are concerned with two classifications of water storage.

1. Raw water storage: Water supplies that are used to feed water to a filtration and treatment plant for purification in order to produce **finished water** that is used for domestic purposes including drinking water is classified as **raw water**. Raw water sources from streams, rivers, ponds, lakes, and even reservoirs are **not suitable** for any domestic purposes including water for cooking, bathing, and especially drinking. The one exception is individual well water that has been chlorinated and disinfected for individual household use in accordance with individual State Public Health regulations.

   Extreme caution: Raw water or any water supply that has not been treated to Environmental Protection Agency (EPA) standards **is not** to be pumped into fire hydrants attached to a municipal water system. During the summer drought conditions of 2005 in the Middle Atlantic States, there were reports where community fire departments were pumping water from creeks and ponds into small water systems because there was no water in the reservoirs to supply the water piping system. This is considered a very dangerous situation, and such practices present serious health risks to persons using these water supplies. Furthermore, all components of the water system are required by either State or Federal regulations to be completely disinfected along with biological testing before the water system can be placed back in service to provide treated water for human consumption.

   A more positive approach to raw water supplies is to use raw water holding basins, ponds, or reservoirs as an alternative water supply source to meet unusual demand on the water system, including a major fire, situations when the main water supply is low, or any other emergency situation requiring large volumes of water such as a primary or secondary water main break. These raw water sources should be arranged so that the water flows by gravity, if possible, to the water treatment plant. If the terrain in the area of the treatment plant does not permit this height differential, then arrangements need to be made for stationary pumps, or even fire department pumpers to pump water from the raw water source to the water treatment plant.

2. Finished water storage: The most common type of water storage on a municipal water system is the use of clear wells on the outboard side of water treatment plants, ground-level water-storage tanks and elevated water-storage tanks to store **finished water** that is suitable for domestic consumption. Therefore, an extremely important element in a water distribution system is **finished water storage**. Water system **storage facilities** have far-reaching effects on a given system's ability to provide adequate consumer consumption plus adequate water supplies for meeting fire-flow demand in addition to consumer consumption. The two common finished water storage methods 1) ground-level storage, and 2) elevated storage, are presented below. Emphasis is placed on the relative merits of both methods.

FUNCTIONS OF DISTRIBUTION STORAGE

Storage within a distribution system enables the system to process water at times when treatment facilities otherwise would be idle. It is then possible to distribute and store water at one or more locations in the service area that are closer to the user.

1. Advantages.

   The principal advantages of distribution storage include the fact that storage equalizes demands on supply sources, production works, and transmission and distribution mains. As a result, the sizes or capacities of these elements need not be so large. Additionally, system flows and pressures are improved and stabilized to better serve the customers throughout the service area. Finally, reserve supplies are provided in the distribution system for emergencies, such as firefighting and power outages.

2. Meeting system demands and required fire flow.

   The location, capacity, and elevation (if in fact elevated) of distribution storage are closely associated with system demands and the variations in demand that occur throughout the day in different parts of the distribution system. System demands can be determined only after a careful analysis of an entire distribution system. However, some general rules may serve as a guide to such analysis. **Table 4-1** lists daily and hourly variations for a typical city and the resultant storage depletion. Such data are of great assistance in determining required storage capacities. However, it should be recognized that each municipal water distribution system has its own specific requirements.

Table 4-1

Water Use and Storage Depletion of Maximum Day in a Typical Municipality

Hour	Ratio of Hourly Demand Rate To Maximum Day Demand Rate	Hourly Variation in Distribution Storage Reserve *mil gal*	Cumulative Storage Depletion *mil gal*
7-8 a.m.	1.00	-0.00	0.00
8-9	1.10	-0.10	0.10
9-10	1.25	-0.25	0.35
10-11	1.28	-0.28	0.63
11-12	1.20	-0.20	0.83
12-1 p.m.	1.18	-0.18	1.01
1-2	1.16	-0.16	1.17
2-3	1.10	-0.10	1.27
3-4	1.00	-0.00	1.27
4-5	1.08	-0.08	1.35
5-6	1.15	-0.15	1.50
6-7	1.30	-0.30	1.80
7-8	1.60	-0.60	2.40
8-9	1.40	-0.40	2.80
9-10	1.25	-0.25	3.05**
10-11	0.90	+0.10	2.95
11-12	0.85	+0.15	2.80
12-1 a.m.	0.70	+0.30	2.50
1-2	0.60	+0.40	2.10
2-3	0.50	+0.50	1.60
3-4	0.50	+0.50	1.10
4-5	0.50	+0.50	0.60
5-6	0.60	+0.40	0.40
6-7	0.80	+0.20	0.20

*Average day, 16 mil gal; maximum day, 25 mil gal; constant hourly supply rate (at maximum day demand rate), 24 mgd or 1 mil gal/h.
**Maximum storage depletion.

Rarely can distribution storage be justified economically in an amount greater than will take care of normal daily variations and provide the needed reserve for fire protection and **minor** emergencies. In systems of moderate size, the amount of water storage available for equalizing water production is 30 to

40 percent of the total storage available for water-pressure equalization purposes and emergency water supply reserves. In normal water system operations, some water from storage should be used each day, not only to maintain uniformity in production and pumping, but also to ensure circulation of the stored water, to prevent ageing of the water which affects water quality. Storage in the distribution system normally is brought to full capacity each night and is increased during low-demand periods of the day.

Normally, it is more advantageous to provide several storage units in different parts of the water distribution system than it is to provide an equivalent capacity at a central location.

Smaller pipelines are required to serve decentralized storage and, other things being equal, a lower flow-line elevation and pumping head result.

ELEVATED AND GROUND-LEVEL STORAGE

Storage within the distribution system normally is provided in one of two ways: elevated storage or ground storage with high-service pumping. It should be noted that elevated storage provides the best, most reliable, and most useful form of storage, particularly for structural fire suppression.

Elevated Storage

Properly sized elevated water tanks provide dedicated fire storage and are used to maintain constant pressure on the water supply distribution system.

Domestic water supplies are regularly fed to the system from the top 10 to 15 feet of water in the elevated tanks. As the water level in the tank drops, the tank controls call for additional high-service pumps to start in order to satisfy the system demand and refill the tanks. The high-service pumps are constant-speed units, which can operate at their highest efficiency point virtually all the time. The remaining water in the tanks (70 to 75 percent) normally is held in reserve as dedicated fire storage. This reserve will feed into the system automatically as the fire-flow demand and the domestic use at a specific time exceed the capacity of the system's high-service pumps.

Ground Storage

Since water kept in ground storage is not under any significant pressure, it must be delivered to the point of use by pumping equipment. This arrangement limits the water distribution system's effectiveness for fire suppression in three ways:

1. There must be sufficient excess pumping capacity to deliver the peak demand for normal uses as well as any fire demand, which requires a generally unused investment in pumping capacity. The pumps are activated periodically to redistribute the water in the holding tank to avoid stagnation of the water.

2. Standby power sources and standby pumping systems must be maintained at all times because the system cannot function without the pumps.

3. The distribution lines to all points in the water distribution system must be significantly oversized to handle peak delivery use plus fire flow, no matter where the fire might occur near one or more fire hydrants on the piping system.

However, in hilly areas it is frequently possible to install ground reservoirs at sufficient elevation so that the water would "float" on the distribution system. This eliminates the need for pumps at the ground-storage facility. If the desired overflow elevation can be achieved on a hill, a considerably larger storage capacity can be installed when compared to an elevated tank. This may result in placement of the storage facility on a hill in a less desirable location. Such a placement would provide larger storage capacity than could be achieved by an elevated storage tank(s), or it should provide the equivalent storage more economically.

When ground-level storage is used in areas of high fire risks, the energy that would be needed to deliver the water is lost on the initial delivery of water to the tank. The water supply must be repumped and repressurized with the consequent addition of more standby generators and more standby pumps. In addition, the system's high-service pumps must be either variable speed or controlled by discharge valves to maintain constant system pressures. This equipment is expensive, uses additional electrical power, and requires extensive operation and maintenance. Frequently, the additional capital costs for pumps, generators, and backup systems, and the long-term energy costs, significantly increase the costs of a ground-storage system.

PUMPING FOR DISTRIBUTION STORAGE

There are two types of water supply distribution storage as defined above:

1. Ground-level storage.
2. Elevated storage.

There also are two types of pumping supply systems. Both of the concepts are expanded upon below. One is a direct pumping system, in which the instantaneous system demand is met by pumping with no elevated storage provided. The second type is an indirect system in which the pumping station lifts water to a reservoir or elevated storage tank, which **floats** on the water system, based on demand, and provides system pressure by the gravity method.

1. Direct pumping.

    The direct pumping system is considered obsolete today, although there are some systems of this type still in existence. Variable-speed pumping units operated off of direct system pressure are also in use in some communities. Hydropneumatic tanks at the pumping station provide some storage capability.

These tanks permit the pumping station pumps to start and stop, based on a variable system pressure preset by controls operating off of the storage tank.

2. Indirect pumping.

In an indirect system, the pumping station is not associated with the demands of the major load center. It is operated from the water level difference in the reservoir or elevated storage tank, enabling the prescribed water level in the tank to be maintained. The majority of systems have an elevated storage tank or a reservoir on high ground floating on the water system. This arrangement permits the pumping station to operate at a uniform rate, with the storage either making up or absorbing the difference between station discharge and system demand.

EVALUATION OF MUNICIPAL WATER STORAGE

Two variations of distribution storage design affect the operation and reliability of a water system's fire suppression capabilities. These two variations involve placement of the storage between the supply point and the major load center or beyond the major load center. A numerical analysis of the following storage designs is presented to provide comparisons and contrasting approaches to the issue of not providing water storage or providing water storage in one of two different design approaches:

- System A: a pumping station to the major center of demand on the water system (load) with **no** elevated storage tank;
- System B: a pumping station to the major center of demand with an elevated storage tank between the supply and the demand point; and
- System C: a pumping station to the major center of water demand with an elevated storage tank beyond the demand point.

1. System criteria.

a. Normal minimum working pressure in the distribution system should be approximately 50 pounds per square inch (psi) and not less than 35 psi during a maximum hour. A normal working pressure in most systems will vary between 50 to 56 psi.

b. Systems must be designed to maintain a minimum pressure of 20 psi at ground level at all points in the water distribution system under fire flow conditions.

c. The maximum daily demand is considered to be 1.5 times the average daily demand.

d. The maximum hourly demand is considered to be 2.25 times the average daily demand.

2.   Model system.

The model system used in the analysis has the following characteristics:

Population = 27,000

Water demand rules:

Average day–27,000x150 gpcd	=	4.0 mgd
Maximum day–4.0x1.5	=	6.0 mgd
Maximum hour–6.0x1.5	=	9.0 mgh
Needed Fire Flow (NFF)=5,000 gpm	=	7.2 mgd
Max. day+fire flow–6.0+7.2	=	13.2 mgd
Max. pressure at the load center	=	50.0 psi
Min. pressure during the max. hour	=	35.0 psi
Min. pressure on max. day+fire flow	=	20.0 psi

System pipelines are all expressed as equivalent lengths of 24-inch pipe with a C factor of 120. Hydraulic gradient is the slope of the line joining the elevations to which water would rise in pipes freely vented and under atmospheric pressure.

System A–No Storage

If no storage is provided in System A as illustrated in **Figure 4-1** at a given demand rate, the pumping station hydraulic gradient must be sufficient to overcome system losses at a demand rate and maintain a minimum of 5 feet of head at the major load center. Thus, the pumping heads required to maintain 115 feet of head, plus the head loss in 40,000 feet of equivalent pipe for the various conditions are as follows:

Demand Rates		**Pumping Head Required**
Average day, 4.0 mgd: 115+(0.67x40)	=	142 ft
Maximum day, 6.0 mgd: 115+(1.42x40)	=	172 ft
Maximum hour, 9.0 mgd: 80+(3.0x40)	=	200 ft
Maximum day, plus fire: 13.2 mgd-46+(6.1x40)	=	290 ft

Figure 4-1

System A–A Hydraulic Gradient With No Storage

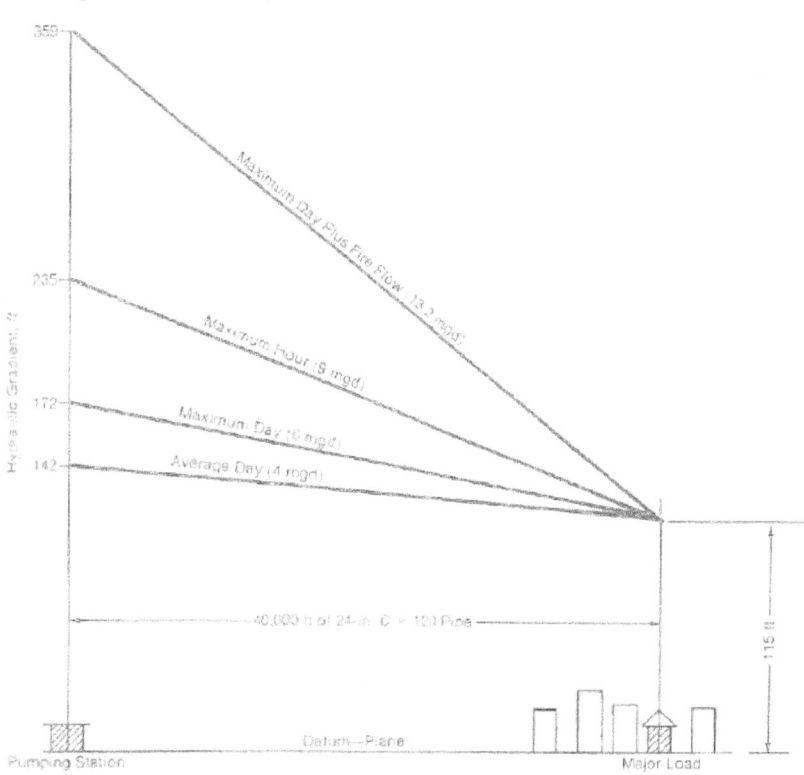

System B–Storage Ahead of Load Center

If, as shown in **Figure 4-2**, a 1-million gallon storage tank is located 130 feet above the datum plane and at a distance of 35,000 feet from the pumping station (5,000 feet ahead of the load center), the pumping head of a given pumping rate must be sufficient to pump against a head at the storage tank and overcome system losses at the pumping rate.

Figure 4-2

System B–Hydraulic Gradients With Storage Between the Pump Station and the Load Center

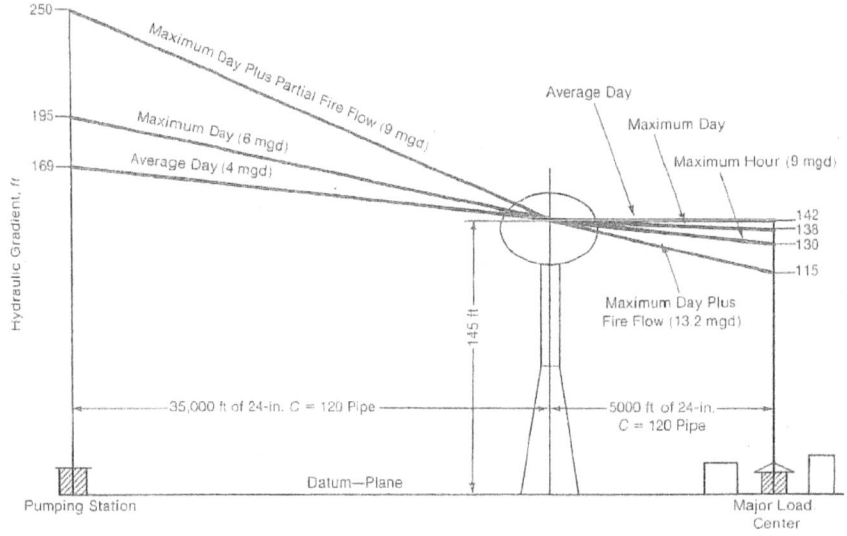

Average day: At the average-day demand, the required pumping rate (no water taken from storage), is 4.0 mgd. The pumping head required is equal to the hydraulic gradient at the tank plus the head loss in 35,000 feet of equivalent pipe at 4.0 mgd, or 130+(0.67x35)=153 ft. The hydraulic gradient at the load center is the hydraulic gradient at the tank minus the head loss in 5,000 feet of equivalent pipe, or 130-(0.67x5)=127 ft.

Maximum day: At the maximum-day demand, the required pumping rate is 6 mgd with no water taken from storage. The pumping head is equal to the hydraulic gradient at the tank plus the head loss in 35,000 feet of equivalent pipe at 6 mgd, or 130+(1.42x35)=180 ft. The hydraulic gradient at the load center is the hydraulic gradient at the tank minus the head loss in 5,000 feet of equivalent pipe at 6.0 mgd, or 130-(1.42x5)=123 ft.

Maximum hour: At the maximum-hour demand, the flow at the load center must be 9.0 mgd and the pressure should get no lower than 35 psi or 80 feet. The pumping head required is equal to the hydraulic gradient at the tank plus the head loss in 35,000 feet of equivalent pipe at the chosen pumping rate. If 1.0 mgd is to be supplied from the tank storage and the remaining 8 mgd is to be supplied from pumping, the pumping head required is 130+(2.41x35)=214 ft as illustrated in **Figure 4-2**.

Maximum day plus fire flow: At the maximum-day demand plus the fire demand, the flow to the load center must be at a 13.2 mgd rate at a minimum pressure of 20 psi, or 46 feet. If 1 mgd is supplied from storage, the pump station would have to pump the remaining 12.2 mgd, and the pumping head required is equal to the hydraulic gradient at the tank plus the head loss in 35,000 feet of equivalent pipe at 12.2 mgd, or 100+(5.2x35)=282 ft.

Demand Rates		Theoretical Pumping Head Required
Average day, 4.0 mgd with no water storage	=	153 ft
Maximum day, 6.0 mgd with no water storage	=	180 ft
Maximum hour, 9.0 mgd; 8.0 mgd rate from pump and 1.0 mgd rate from storage	=	214 ft
Maximum day plus fire flow, 13.2 mgd; 12.2 mgd rate from pumps plus 1.0 mgd rate from storage	=	282 ft

The previous example is theoretical because it assumes that all the demand is at the major load center, which, of course is not the case in an actual community.

System C–Storage Beyond Load Center

In the arrangement shown in **Figure 4-3**, 1-million gallons of water storage is provided 5,000 feet beyond the load center (45,000 feet from the pump station) at an elevation of 119 feet above the datum plane. When no water is being taken from storage at a given demand rate, the pumping head must be sufficient to pump against the head at the tank and overcome losses between the pump station and the load center at the demand rate. When part of the demand is being supplied from storage, however, the pumping head only needs to be sufficient to pump against the head at the load center and overcome losses at the pipeline between the pump station and the load center.

Figure 4-3

System C–Hydraulic Gradients With Storage Beyond Load Center

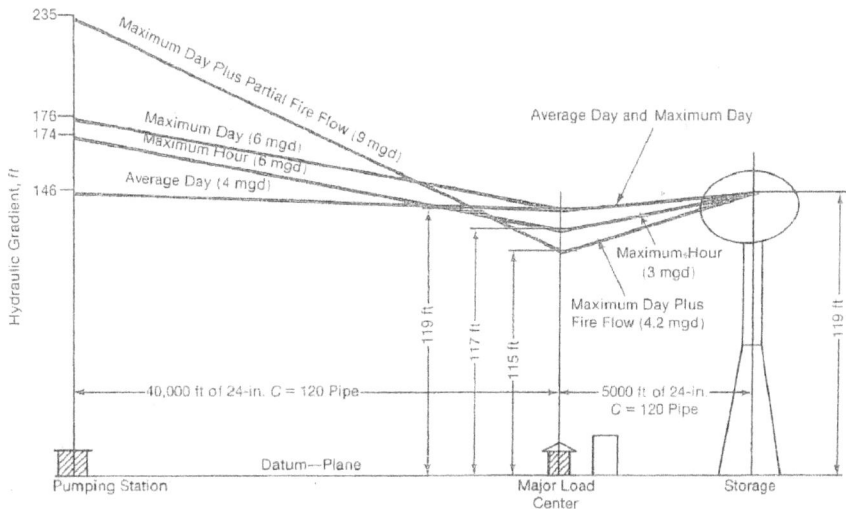

Average day: At the average-day demand, the required pumping rate is 6 mgd (no water taken from storage). The pumping head required is equal to the hydraulic gradient at the tank plus the head loss in 40,000 feet of equivalent pipe, or 119+(0.67x40)=146 ft. The hydraulic gradient at the load center is thus identical to that at the tank (119 ft).

Maximum day: At the maximum-day demand, the required pumping rate is 6 mgd (no water taken from storage). The pumping head required is equal to the hydraulic gradient at the tank plus the head loss in 40,000 feet of equivalent pipe at 6 mgd, or 119+(1.42x40)=176 ft. The hydraulic gradient at the load center is identical to that at the tank (119 ft).

Maximum hour: For a maximum-hour demand (9 mgd rate), the working pressure can be lowered to 35 psi or 80 feet. The storage tank could supply water at a rate of 700 gpm (1.0 mgd) during the hour and the remaining 8.0 mgd rate would be supplied from pumping. The hydraulic gradient at the load center is the hydraulic gradient at the tank minus the head loss in the 5,000 feet of pipe between the tank and the load center at the storage discharge rate of 1.0 mgd, or 119-(0.05x5)=119 ft. The pumping head required is equal to the hydraulic gradient at the load center plus the head loss in 40,000 ft. of equivalent pipe at 8.0 mgd, 119+(2.41x40)=215 ft.

Maximum day plus fire flow: The pressure at the load center can be lowered to 20 psi or 46 feet during a fire. Assuming the storage tank is full (1.0 mgd) this flow can move from the tank to the load center at a 2,800 gpm (4.0 mgd) rate. Thus the remainder of the demand (9 mgd) must be supplied from pumping. The pumping head required is equal to the hydraulic gradient at the load center (46 feet) plus the head loss in 40,000 feet of equivalent pipe, or 46+(3x40)=166 ft.

Demand Rates		Theoretical Pumping Head Required
Average day, 4.0 mgd with no water from storage	=	146 ft
Maximum day, 60 mgd with no water from storage	=	176 ft
Maximum hour, 9.0 mgd; 8.0 mgd rate from pump plus 1.0 mgd from tank	=	215 ft
Maximum day plus fire flow, 13.2 mgd; 9.2 mgd rate from pumps plus 4.0 mgd from tank	=	166 ft

COMPARISON OF SYSTEM A WITH SYSTEM C

If no storage is provided, 124 ft (290 ft-166 ft), more pumping head is required to furnish the maximum-day demand plus fire flow than if adequate storage is provided beyond the load center. With the increased pumping rates required with no storage, the power needed is approximately 1,100 horsepower (hp), as opposed to 495 hp with storage, or more than twice as much. Similarly, furnishing the maximum-hour demand without storage would require 500 hp, as opposed to 245 hp, which is still more than twice as much.

The capacities of the pumps required under these two conditions would be 13.2 mgd at 359 feet of head, as opposed to 9.0 mgd 235 feet of head, and 9 mgd at 235 feet of head, as opposed to 5.0 mgd at 174 feet of head. During average and maximum day demand, the pumping head at the source of supply is approximately the same.

COMPARISON OF SYSTEM B WITH SYSTEM C

In comparing storage located between the source and the load center with storage located beyond the load center, the example illustrates that pumping heads at the pumps are substantially lower under all conditions if the load center is located between the pumping station and elevated storage.

Recommended Water Supply System Design Practice

The American Water Works Association (AWWA), *Manual of Water Supply Practices*, AWWA–M-31 states that the recommended water distribution system design is to have the major load center located between the pumping station and an elevated water storage tank of sufficient capacity. This is the most cost-effective design from a capital cost and operating cost standpoint. (Reference #1, below)

References:

1. American Water Works Association. *Distribution System Requirements For Fire Protection*, AWWA–M-31. Denver: Author, 1992, pg. 31.

2. Cote, A.E., and J.L. Linville, Eds. *Fire Protection Handbook*. 16th Ed., Quincy: National Fire Protection Association, 1986.

3. Fair, G.M., et al. *Water and Wastewater Engineering*. New York: John Wiley and Sons, Inc., 1976.

4. Steel, E.W., and T.G. McGhee. *Water Supply and Sewage*. New York: McGraw-Hill Book Company, 1979.

Chapter 5: A Comprehensive Method For Evaluating Municipal Water Supply Delivery Systems

Scope and Content

The evaluation of municipal water systems starts with a through evaluation of the water system source or sources, extends through the treatment of raw water to provide finished water or purified water for consumer consumption, moving this water through distribution piping to provide consumer consumption on demand and Needed Fire Flows (NFFs) on demand to confine, control, and extinguish developing fires in structures. Of no less importance is determining just how well the municipal water system **measures up** in accomplishing this task 24 hours a day, 365 days a year, year in and year out. Each component or element of the water system can be analyzed individually to carry out such an evaluation, or the water system can be evaluated as a whole, integrated system to understand the adequacy and reliability of the **complete** water supply delivery system for all service areas of a specific community.

The Insurance Services Office, Inc. (ISO), uses the *Fire Suppression Rating Schedule* (FSRS) as one tool in assessing risk levels for subscribing insurance carriers throughout the United States. One significant item in the FSRS covers municipal water supply in a systematic and comprehensive approach. Municipal officials, water supply officials, and fire service officials **need to know** how the insurance industry evaluates municipal water supplies. These evaluations conducted approximately every 10 years by ISO field representatives culminate in a published report that may be requested through the regional office of the ISO by the chief executive of the of the community or graded area, (i.e., county, city, community, village, town, township, or similar political jurisdiction).

This chapter covers both general and specific evaluation criteria on municipal water systems that can be applied to any given installed water system. This information provides a very important guideline for evaluating a water delivery distribution system systematically for both consumer consumption and fire flow **availability**. It also provides quantitative measures on just how well the water supply system and the water distribution system are meeting the needs of the population for purified water and the needs of the community for fire protection water to confine, control, and extinguish developing structural fires. This information helps community leaders to build a profile of the existing water delivery system and identify both the strengths and the weaknesses. Maybe most important, it calls **attention** to a large number of topics that need to be considered in the evaluation of any given water supply system and the distribution system to supply finished water to consumers while making available NFFs for adequate fire protection.

GENERAL CONSIDERATIONS

Property and casualty insurance underwriters base their business on the level of risk they are assuming in writing insurance policies. In relation to fire protection, they need to know about the adequacy and reliability of both public sector and private sector water supplies that are suitable and in place for fire extinguishment. From this perspective, **adequacy** is considered the ability of a water system to supply the NFF at identified locations throughout a community (i.e., county, city, village, township, etc.). In other words, available fire flow should equal or exceed NFF for any area of selected commercial, industrial, and residential property type risk.

Using criteria documented in the ISO's FSRS (latest edition, 2003) a water supply evaluation is conducted by field representatives for the ISO. Section I of the *Schedule* considers properties that have a NFF below 3,500 gallons per minute (gpm). Section II considers properties that have a NFF above 3,500. The community grading evaluation is limited to Section I of the FSRS. Individual properties with fire flows in excess of 3,500 gpm are evaluated on an individual basis for insurance underwriting purposes. Therefore the community as a whole is not penalized for deficiencies in meeting the water supply demands for large-scale fire risks. Furthermore, it needs to be recognized that fire flow durations at representative fire sites are evaluated as required for 2, 3, and 4 hours; this is a significant consideration for any public sector water system.

A *Grading Schedule* water supply evaluation considers three primary components of the water system:

1. The supply works capacity.

2. The water main capacity.

3. The distribution of fire hydrants.

Based on these three considerations, the capacity of the water system is evaluated at each test location. Water supply is credited at each test location based on the rate of flow in gpm at 20 pounds per square inch (psi) residual pressure. The **lowest** rate of flow is used for credit based on the following factors:

- NFF in gpm;
- capacity of the supply works in gpm;
- capacity of the water supply distribution system in gpm as determined from fire hydrant flow tests; or
- the combined capacity of water delivery from all fire hydrants located with 1,000 feet of the NFF site.

The objective of any water supply evaluation is to determine if a given water system is capable of supplying the NFF at each selected representative fire-risk site throughout the community. When the delivery capacity is less than the NFF, it is important to determine which water system component is limiting the available fire flow. In the following sections, several general topic headings concerning water supply are reviewed before examining specific components of the supply system.

PART OF THE CITY/COMMUNITY UNPROTECTED

For the purpose of water supply evaluation using criteria in the ISO FSRS, property within 1,000 feet of a recognized fire hydrant is considered to be **protected**. This should not be confused with the term **protected property** as used in fire insurance rating making, where the term is used to designate property that is a protected by automatic sprinkler systems.

A **recognized fire hydrant** has a minimum supply capability of 250 gpm at 20 psi residual pressure for a 2-hour duration.

A community grading evaluation carefully considers all built areas of the community that are unprotected because of inadequate water supply capability. These areas are beyond 1,000 feet of a recognized fire hydrant. The 1,000 feet is measured as hose would be laid from a fire hydrant to the property site by fire apparatus. Unprotected areas, defined by the lack of a recognized water supply, generally receive a semiprotected classification of 9, which indicates that the property owner(s) will pay a higher insurance premium than property located in a protected area.

MAXIMUM DAILY CONSUMPTION RATE

As previous identified, a municipal water system is constantly providing domestic water for consumption by industry, business, and individuals. This same water system needs to provide water for structural fire suppression on demand, which can occur at any time of the day, day of the week, or during any month of the year. It is not only reasonable, but it occurs in the real world, that serious fires erupt in a community when water usage for domestic services is at the highest level. Therefore, a water system should be capable of delivering a NFF during periods of maximum consumption. This is defined in the *Grading Schedule* as Maximum Daily Consumption (MDC), which is determined from water department records. In the absence of complete and accurate records, an ISO Field Representative may estimate the MDC rate by using 150 percent of the Average Daily Consumption (ADC). In 2004, the American Water Works Association (AWWA) reported that the national average water consumption was 148 gallons per day per person. (Reference #1, pg. 99)

Example 1: Determine the estimated ADC and the MDC for a community of 12,500 people using 148 gallons per day per person as noted above.

- ADC=12,500x148 gal/person=1,850,000 gallons of water per day.

- The ADC rate per minute=1,850,000 gal divided by (24 hours in a day x 60 minutes in an hour)=1,285 gpm.

- The MDC rate per minute can be estimated from the ADC rate.

- MDC rate=1,285 gpm(ADC)x150 percent=1,927 gpm.

This type of estimation can be used in the absence of a good hourly water consumption graph record on the water supply system.

Minimum Pressure

The ISO *Grading Schedule* specifies that a water system is reviewed at a residual pressure of 20 psi. (FSRS p. 25)

Flow tests are conducted on fire hydrants in the vicinity of representative fire risks in a community. To determine static pressure, (no flow condition), residual hydrant pressure, or flowing pressure, and the calculated flow in gpm. Test results can be used to compute the flow available at the residual pressure fire hydrant in gpm when the residual pressure remains at 20 psi. (Refer to the graphing methods presented in Chapter 3.)

The 20 psi value is used to avoid cavitation in water mains or suction hose between the fire hydrant and the connected fire department pumper.

Needed Fire Flow and Fire Flow Duration

The *Grading Schedule* states that the fire flow duration should be 2 hours for NFF up to 2,500 gpm, and 3 hours for NFF of 3,000 to 3,500 gpm. However, water supply capability also affects the community evaluation for individual property fire suppression where the identified NFF exceeds 3,500 gpm as covered under Section II of the FSRS (i.e., *Grading Schedule*). Under Section II, it is specified that individual buildings having a NFF greater than 3,500 gpm, but not exceeding 12,000 gpm, should have a 4-hour fire-flow duration.

Key Consideration: It is very important to understand that flow duration is based on the NFF **and** the MDC rate at each selected test site.

SERVICE LEVEL

The *Grading Schedule* indicates that a service level is a part of the city water distribution system, which is served by one or more sources of supply, but which is separated from the remaining distribution system by closed valves, check valves, or pressure-regulating equipment, or is not physically connected as part of the main system. The need for more than one service level in a community is based on one or more of the following conditions:

⬥ Hills and valleys can cause topography changes of a magnitude that require different pressure zones on a water system to prevent excessive pressures.

⬥ Portions of communities often have natural or manmade barriers, such as rivers and railroad yards, the require zoning the water system.

⬥ Portions of the water system may be supplied from different water sources. This could include one supply from a well system and one system that is fed by gravity or standpipe tanks.

Figure 5-1 illustrates one common type of water system with two pressure zones. A gravity-feed reservoir supplies approximately 80 percent of the community. A substantial residential area and a shopping center are located on a plateau at a point remote from the water source. A pumping station (illustrated) is required to boost water pressure to the plateau area. Hydraulically, the pumping station establishes a second zone for the water system. The static and residual profile for the built-up area on the plateau is different from the profile for the central city.

Figure 5-1

Water System With Two Pressure Zones

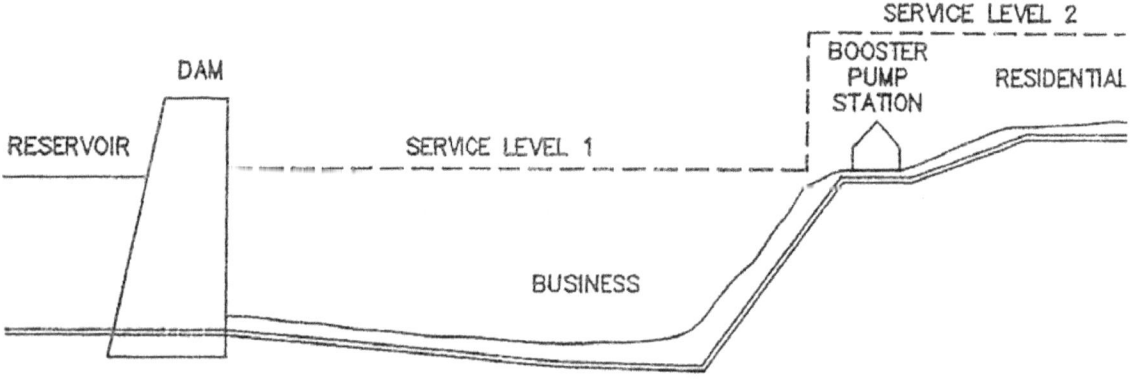

The *Grading Schedule* indicates that, when a water system is supplied from two or more sources or supply works, the credit is to be based upon the combined protection provided from all sources or supply works. This concept is extremely important when considering options for meeting both consumer consumption demand and NFF demand. In other words, more than one supply source can be used to meet the water supply demand at specific fire risks. One very important approach is to use alternative water supplies to make up the difference between available fire flows and NFFs.

REVIEW OF THE WATER SUPPLY

Basic criteria for the review and analysis of water systems are found in a number of applicable AWWA publications. The ISO–Commercial Risk Services uses the AWWA criteria as the foundation for evaluating municipal water systems for fire protection. A selected list of important AWWA publications on water supply systems is provided in the reference section. These references, along with the ISO *Grading Schedule*, are used in the systematic evaluation of community water systems as presented below. For each representative fire-risk location in a community, the supply works, water mains, and fire hydrant distribution are reviewed for **actual** water delivery capability to that site. This is accomplished by studying water supply records and conducting fire-flow tests.

SUPPLY WORKS

A water system supply works evaluation examines the amount of potable water that can be delivered to the distribution system piping, often called the water mains. This evaluation considers a number of factors that affect the supply capacity:

- minimum storage of water;
- municipal water supply pumps;
- water filters and treatment facilities to provide potable water; and
- emergency water supplies.

Two special topics, Water System Supplies and Fire Department Supplies also are evaluated here. These topics are of importance to all communities, especially those that have relatively poor water delivery capability, or cities that have areas where the available water supply does not meet NFF. Furthermore, these two topics are most important for fire protection to areas that are beyond 1,000 feet of a recognized fire hydrant (i.e., a fire hydrant that delivers a minimum of 250 gpm for a 2-hour duration).

As discussed under Alternative Water Supplies in Volume I, the *Grading Schedule* provides a means of obtaining credit for water supply at specific fire risks when a fire department can demonstrate a minimum water supply delivery capability of at least 250 gpm for 2 hours. Selected water sources can include bays, rivers, canals, streams, ponds, well, cisterns, or other similar sources.

WATER SYSTEM SUPPLY COMPONENT

It is necessary to understand the types of community water supply systems and the arrangement of components prior to discussing the evaluation concepts associated with them. There are two basic types of water supply systems:

1. Gravity-feed system.
2. Pressure-feed system, where pressure is developed by stationary pumps.

Each of these basic types of supply systems can be subdivided according to the type of potable water and nonpotable water in storage:

◆ reservoirs that hold nonpotable water for gravity feed;

◆ elevated tanks, standpipe tanks, and impounding tanks that store potable or finished water for gravity feed;

◆ pumping stations that are supplied by ground-water sources: lake, river, ponds, etc.; and

◆ pump(s) at well sites.

In any of the above systems, pumps may be used to pump nonpotable water to a filtration system or purifying system, followed by pumping potable water to holding tanks or directly into the water supply distribution system. The figures below depict the basic types of water supply systems. Some of these systems will be used in the evaluation examples that follow.

Figure 5-2

Gravity Reservoir System

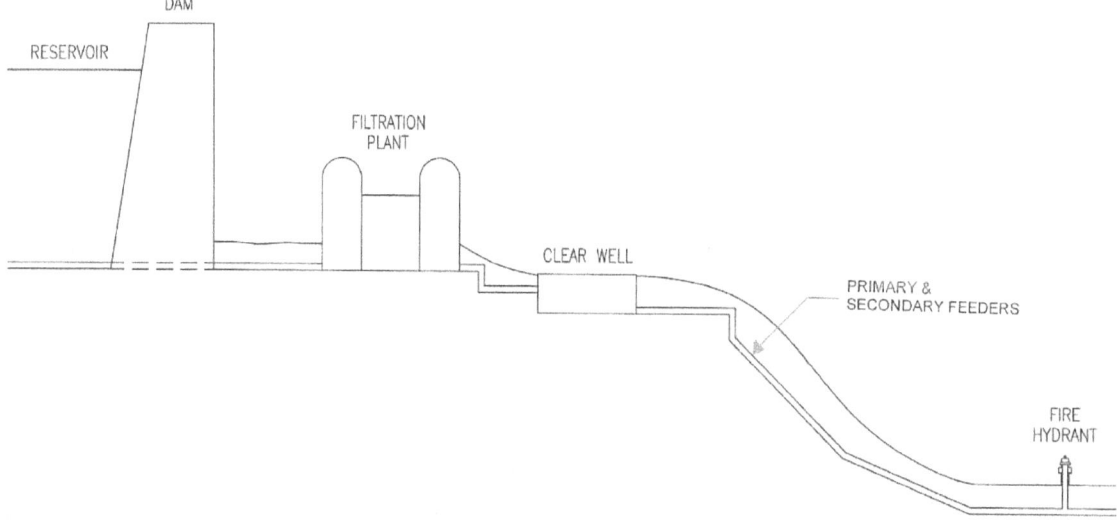

Figure 5-2: Gravity Reservoir System

An impounding reservoir is sown at the left of the figure. Untreated water flows under the dam to a filtration plant. Water passes through filter beds to an additive metering system to chlorinate the water and control the pH factor. Other water treatment processes may be necessary, based on the quality of water after this preliminary and standard treatment process. Once the water passes potability tests to meet the Environmental Protection Agency (EPA) drinking water standards, it then flows to a clear well holding tank or smaller reservoir. Water from the holding area flows into the distribution system by gravity feed.

A gravity reservoir system is considered the most reliable type of the water systems, since all of the water flows by gravity, and there is no mechanical interface.

Figure 5-3

Pumping Station and Gravity Feed

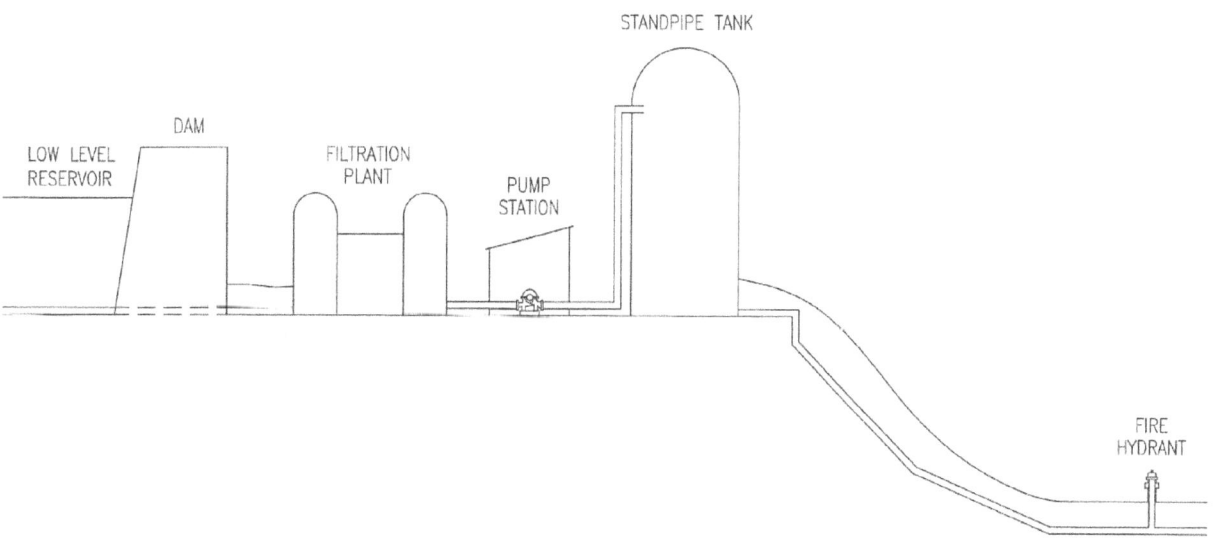

Figure 5-3: Pumping Station and Gravity Feed

In this illustration, domestic water flows from a low-head pressure source to a pumping station. The water is pumped to a filtration plant. After filtration and purification to meet the EPA drinking water standards, water is pumped to a 500,000-gallon gravity standpipe tank. Water flows to the distribution system on demand from the storage tank. Altitude valves are used to control the purified water and activate stationary pumps to maintain water in the standpipe tank. This system combines the reliability of a gravity feed to the distribution water supply network with a pump station to supply the water from a static source through to the finished water in storage for delivery for consumer consumption and fire demand.

Figure 5-4

Pumping Station at Well Site(s) and Gravity Storage

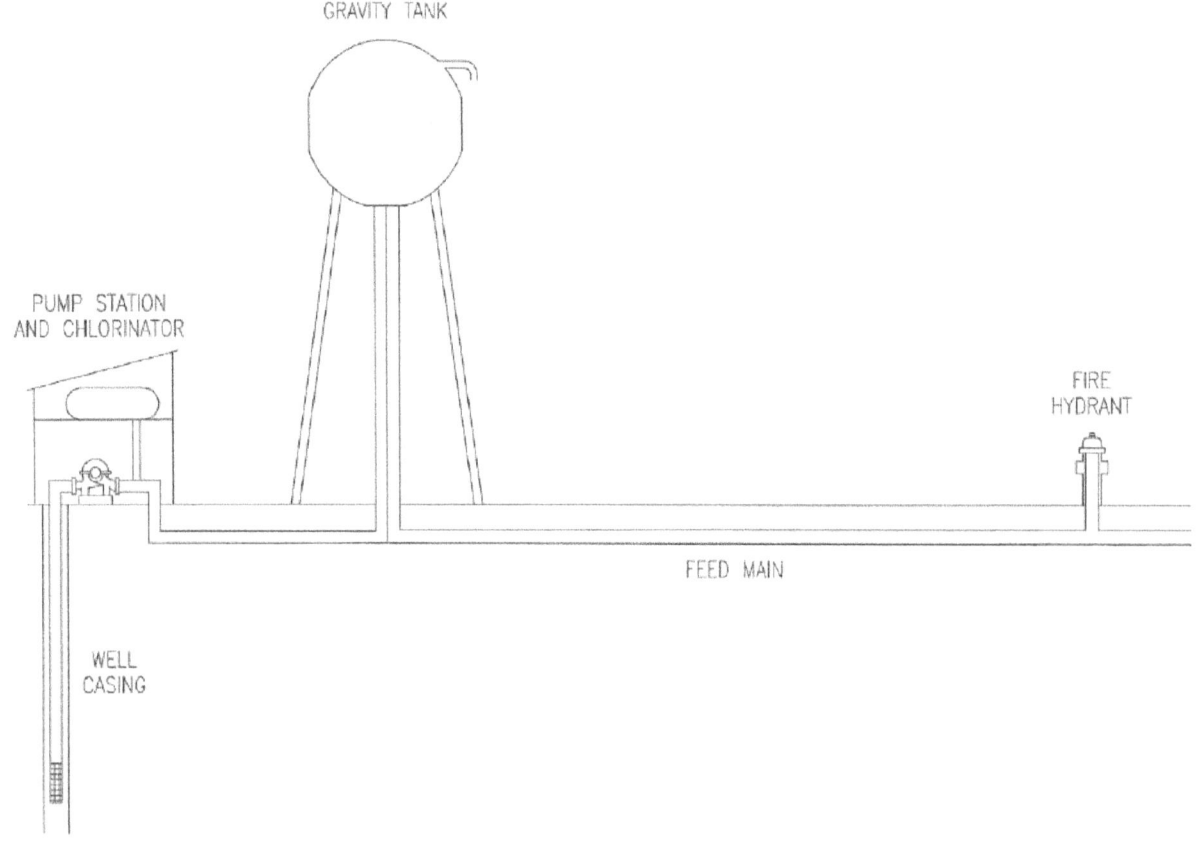

Figure 5-4: Pumping Station at Well Site(s) and Gravity Storage

One well, or a field of wells, feeds water to a ground-level pumping station as illustrated. Well water typically is treated by chlorination through an injection method. Any other required water treatment generally is handled in the same manner. Treated water is pumped to an elevated storage tank. Water flows from the storage tank to the distribution system.

Figure 5-5

Direct Pumping System

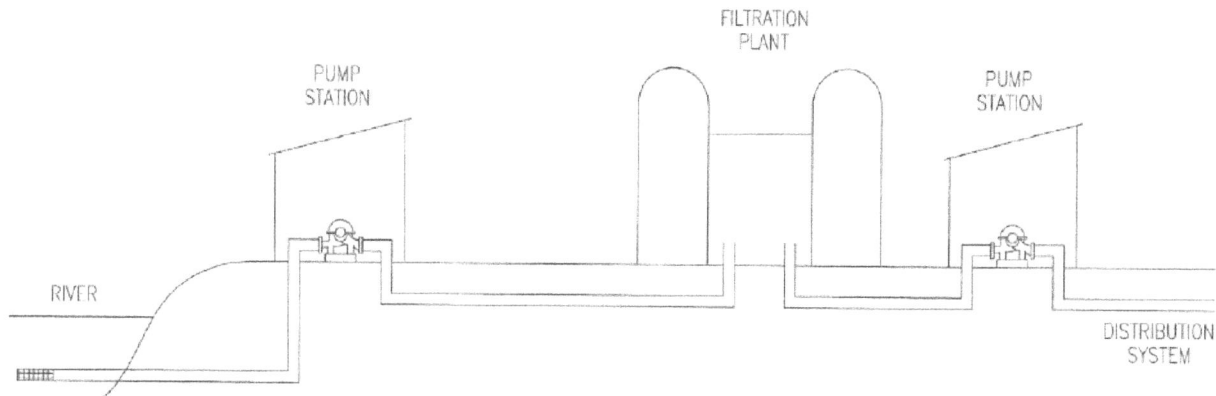

Figure 5-5: Direct Pumping System

This figure shows how a direct pumping station feeds water from a river to the distribution system. The principal feature of this system is that all water is transported through pipe and filtering systems by pumping. One or more pumps are on line at all times. Typically, additional pumps are brought on line to meet fire-flow demand. Adequacy and reliability of the water system generally are dependent on the pumping system.

MINIMUM STORAGE

Minimum storage available for public fire protection should be examined in terms of the type of distribution system. It may be helpful to refer to the figures illustrating basic types of water systems. Every water system for communities over 50 population has to have a water storage component in order to provide even minimum fire protection water supplies

On this matter, the *Grading Schedule* states that the average daily **minimum** water storage maintained is the **maximum** amount that can be credited. This concept can be understood through the following explanation.

The amount of water in storage for a given distribution system is constantly changing due to residential, business, and industrial consumption and, as needed, fire-flow consumption. As previously identified, consumption varies by the time of the day and the day of the week. Consequently, finished water (referring to water that has been suitably treated to meet EPA Safe Drinking Water Standards), in storage also varies by the time of day and the day of the week. The insurance community grading process is interested in the **average** minimum storage as a benchmark for both water system adequacy and reliability.

To determine minimum storage, it is necessary to maintain **accurate** records on storage facilities including clear wells, standpipe tanks, and gravity tanks. Chart recorders linked to each storage facility typically are used to cover the capacity range of the specific storage tank. The recorded minimum amount of water in storage for each day is expressed in gallons or millions of gallons. For each year being evaluated, all daily minimum amounts are totaled and divided by 365 days to provide the average minimum water in storage. This is the amount of water credited as available for consumer consumption and fire flow at any given time. Obviously, consumer consumption does not stop when there is a fire in the community.

The ISO *Grading Schedule* does specify some conditions and exceptions to the concept of minimum storage:

◆ The absolute minimum water supply under extreme dry weather conditions should not be considered; this is a judgment item.

◆ Only the portion of water in storage that can be delivered at the required pressure of 20 psi at representative tests sites is to be considered.

◆ Water loss due to a pipe rupture should not be considered in the evaluation.

◆ Water supply available during periods when water tanks are being cleaned and painted has to be carefully evaluated by an ISO field representative.

It should be recognized that direct pumping systems from treated water supplies, or nonpotable water that is treated during the pumping process, generally are limited by both water storage and pumping capacity. The pumping limitation is discussed below.

MUNICIPAL STATIONARY PUMPS

Municipal pumps may interface with a water distribution system supply in several ways. Some of these ways are illustrated above. The most common uses of pumps at the supply works follow:

◆ Pumps take water from an impounding source (lakes, rivers, streams, etc.) and send the water through pipes to a filtration/treatment plant.

◆ Pumps take treated water from clear wells and pump directly into the water delivery system piping network.

◆ Pumps take treated water from clear wells and pump to elevated or standpipe tanks.

◆ Pumps take water from wells and pump it through purifiers and then to the water distribution system.

The *Grading Schedule* indicates that stationary municipal pumps should be credited at their effective capacities when delivering at normal operating pressures as specified by the pump manufacturer. This information needs to be taken from the manufacturer's pump specification plate and manufacturer's pump curves. This information is to be available at the water supply works. Each water supply pump needs to be tested annually to match points on the manufacturer's pump curve with actual flow capability.

The **actual** flow capability of one or more pumps at the supply works may be limited by the following factors:

- size and length of the suction pipe to the pump;
- filter arrangements;
- Venturi fittings for chlorinators, water softeners, fluorides, and other additives;
- head loss characteristics on the discharge side of the pumps; and
- minimum stored water capacity may limit pump capability when pumping from this source.

The following explanatory information should be of assistance in the determination of creditable pump capacity:

- The total pumping capacity should be the sum of **all** pump facilities available at the test location expressed in gpm.

- When there are two or more pump lifts in a series, the effective pump capacity is the capacity of the lift with the lowest total capacity.

- When the same pumps can operate in two or more lifts, they are to be credited in each lift to determine the lift with the lowest total capacity.

FILTERS

An arrangement of water filters may be used to purify water to meet EPA or public-health standards at the State level. Water passes through filters or filter beds to produce finished water that is now suitable for the distribution system of pipes to supply domestic consumers.

The capacity of filters in gpm or million gallons per day is used to credit filter capacity in the *Grading Schedule* based on the following distribution conditions:

💧 The total filter capacity of each pressure filter or filter bed is to be the sum of the filter capacities expressed in gpm when the filter arrangement discharges directly into the distribution system.

💧 Filters may be considered as operating at reasonable overload capacity based on documented records.

💧 One or more filter system may be limited by pump capacities and pump arrangements.

EMERGENCY SUPPLIES

The *Grading Schedule* considers the ability of the community being evaluated to use emergency water supplies. Emergency water supplies can include, but are not limited to:

💧 cross-connections to adjacent city water systems;
💧 cross-connections to industrial, military, or other government water systems; and
💧 Separate water storage that normally is not used on the distribution system.

The potability of separate water sources has to be considered. Only water that moves through water mains and meets health regulations can be credited as an emergency water supply.

It should be observed that the *Grading Schedule* is essentially concerned with potable water systems that provide water for consumer consumption, plus water for fire protection supplied by fire hydrants. However, the *Grading Schedule* provides an opportunity to supplement existing fire flow at specific risk sites. These sources of supplies include alternative water supplies and other options presented in both Volume I and Volume II.

SUPPLY WORKS CAPACITY

This item provides ISO field representatives with the method for calculating credit for supply works. Consideration is given to each of the flow durations specified in the *Grading Schedule*:

💧 NFF of 2,500 gpm or less   = 2 hours
💧 NFF of 3,000 and 3,500 gpm   = 3 hours
💧 NFF of 4,000 gpm or higher   = 4 hours

While water supply evaluations consider delivery at each representative test location, primary consideration is given to the water supply that enters the water distribution system, because this water is considered to be available instantaneously when the first-pumping fire apparatus arrives at the fire site and connects to a fire hydrant. Alternative water supplies and optional water supplies may result in a time delay before water actually can be delivered at the fire site.

ADMINISTRATIVE PERSPECTIVE ON WATER SUPPLIES

The ISO FSRS only considers the adequacy of the water supply works, not supply reliability. The impact of a water main break, pumps, out of service, filter beds down, etc., is not part of the *Grading Schedule* evaluation. However, municipal officials have a responsibility to consider water systems **reliability** carefully because of the economic impact it has on a community if there is no water supply. Furthermore, insurance rates are based on the assumption that there is a minimum of 500 gpm from the nearest fire hydrant to the insured building. The failure of a water supply at the time of a fire **could** jeopardize insurance payments to building policyholders, or result in lawsuits against the community that is responsible for water supply to individual properties.

The reliability of a community water distribution system can be increased by adding finished water storage capacity. The storage can be arranged to flow into the water system according to consumer consumption. However, sensing valves can be used to hold a specified minimum amount of water for an emergency condition such as a city fire or an outage of some water system component.

Cross-connection to adjacent water supply systems where possible increases both the adequacy and reliability of a given community water system. The interface capability of water supply systems (water districts) should be determined carefully. There are counties now that are providing large water main interconnectors that provide cross-feeders for several communities within the county and water supplies to such buildings as schools throughout a given county. A written agreement between cities to share water supply capabilities under emergency conditions provides benefits to both recipients of the agreement, and potentially improves the insurance grading evaluation for each water district.

It is essential that the minimum amount of water storage for a given city be considered carefully and examined annually.

EVALUATING WATER MAIN CAPACITY

As previous documented in Chapter 1, the following elements comprise a water distribution system used to supply water to commerce, industry, residences, and fire hydrants:

- primary distribution pipe heading from the supply works;
- secondary feeder pipe looping around the major sections of the city;

- distribution pipe laid along individual streets that should interconnect with the secondary feeders; and
- distribution pipe laid along individual streets.

The flow capability, or hydraulic characteristic, of water mains is used to determine the amount of water available at fire hydrants located near NFF sites. Fire-flow tests are conducted on one, or a set, of fire hydrants to measure the flow at a 20 psi residual pressure. These measured flows can be compared to NFFs at the same locations.

Water tests need to be conducted and recorded for each NFF risk site. Both the NFF and the actual fire flow (AFF) tests are conducted in accordance with Chapter 3.

The *Grading Schedule* adds the following instruction:

> If tests are made on two or more systems or service levels at the same location, credit will be given for the sum of the test results on each system, or service, up to the limit of the supply, for the flow duration.

The concept of available fire flow for **different** durations will be covered in the following review.

EVALUATING FIRE HYDRANT DISTRIBUTION

The *Grading Schedule* provides the following information (that is not directly quoted): A review is conducted at each fire hydrant within 1,000 feet of a representative test site location (i.e., fire risk) measured as fire hose can be laid by fire apparatus in order to satisfy the determined NFF.

Proximity of fire hydrant distribution to NFF sites or fire-risk sites, is the third factor in determining water system capability. Credit for fire hydrants is expressed in gpm, based on measured distance from the building site as established above. The flow and distance relationship is as follows:

Credit Up To:	Distance From the Risk Site
1,000 gpm	Within 300 ft of site location
670 gpm	Within 301 ft to 600 ft of site location
250 gpm	Within 601 ft to 1,000 ft of site location

The maximum credit for a fire hydrant is to be limited by the number and size of the outlets as follows:

Hydrant Outlets	Maximum Credit
At least one pumper outlet	1,000 gpm
Two or more hose outlets	750 gpm
One hose outlet only	500 gpm

EXAMPLE ON EVALUATING FIRE HYDRANT DISTRIBUTION

A NFF risk site has four functional fire hydrants that flow at least 250 gpm, each within 1,000 feet of the fire risk site. The credit for these hydrants is as follows:

[Sidebar: The graphical technique presented in Chapter 3 is used for the following determinations.]

A. List each fire hydrant with the creditable flow:

1) Hydrant A   = 750 gpm
2) Hydrant B   = 1,000 gpm
3) Hydrant C   = 670 gpm
4) Hydrant D   = 250 gpm

B. Total the potential flow capability for all four fire hydrants:
The summation is 750+1,000+670=250=2,670 gpm

EVALUATING CAPABILITY OF THE WATER SYSTEM AT EACH TEST LOCATION

The concept is to identify through calculations for NFF, supply works capacity, water main capacity, or hydrant capacity, based on distribution to identify the factor that limits water supply to a selected building site. The creditable rate of flow at each test location is the **lowest** of each consideration. This can be observed from the following example for a hardware store in Yourville. The following values were computed:

A. NFF   = 3,000 gpm
B. Supply work capacity   = 4,000 gpm
C. Water main capacity   = 2,500 gpm

In this example, the water delivery system is limited by water supply tests which demonstrate that the main capacity is 2,500 gpm which becomes the credited value for the hardware store.

CREDIT FOR THE SUPPLY SYSTEM

Credit for available fire flow throughout the community is summarized in by the following formula:

CSS=TLCx35

Where:

CSS=Credit for supply system

TLC=A summation of the creditable fire flow at each test location

The computation is accomplished in the following manner using the constructed Table:

Test No.	Needed Fire Flow	Credited Fire Flow
#1	2,000 gpm	1,500 gpm
#2	3,000 gpm	3,000 gpm
#3	1,500 gpm	500 gpm
#4	2,000 gpm	1,000 gpm
#5	500 gpm	750 gpm
#6	500 gpm	250 gpm
Total	**9,500 gpm**	**7,000 gpm**

Basic Credit Analysis:

Credited fire flow	= 7,000 gpm
NFF	= 9,500 gpm
Credit for the supply system	= 7,000/9,500x35
Credit for the supply system	= 25.8 percent

Using the condition of the example above, this same community earns 25.8 percent out of 35 percent credit for water supply, distribution mains, and fire hydrant distribution for this stand-alone item is 73.7 percent adequate.

FIRE HYDRANTS: SIZE, TYPE, AND INSTALLATION

The *Grading Schedule* indicates that fire hydrant type and installation are evaluated by an earned point system. This is illustrated as follows:

Points are prorated from the following subitems, according to the number of hydrants of each type in the community, compared with the total number of installed fire hydrants as follows:

	Points
With a 6-inch or larger branch, a pumper outlet, with or without 2-1/2-inch outlets	100
With a 6-inch or larger branch, no pumper outlet, two or more 2-1/2-inch outlets, or with a small foot valve or a small barrel	75
With only one 2-1/2-inch outlet	25
With less than a 6-inch branch line	25
All flush-type fire hydrants	25
Cistern or suction point	25

Deduct 10 points if more than one thread is used for pumper or hose outlets.

Deduct 2 points for each 10 percent of the hydrants not opening in the direction of the majority, or with operating nuts different from the majority.

Maximum points for fire hydrants as specified is 100.

The fire hydrant evaluation considers the following. First, it is unlikely that all fire hydrants in a community are of the same type with the same lateral or branch connection. Therefore, it is necessary to compute the point credit for fire hydrants based on the number in each of the above categories. This is accomplished as follows for the above community under consideration.

The community has a total of 1,702 operating fire hydrants. The number in each category follows:

- Number with a 6-inch branch and a pumper outlet   = 450
- Number with a 6-inch branch and no pumper outlet   = 391
- Number with a 4-inch branch or lateral   = 861

To assign a point value it is necessary to convert each of the above number counts into percentages:

- 450/1,702   = 26 percent

♦ 391/1,702   = 23 percent

♦ 861/1,702   = 51 percent

Each calculated percent is multiplied by the point value associated with the type of fire hydrant as follows:

♦ 26 percent x 100 points   = 26 points

♦ 23 percent x 75 points   = 17 points

♦ 51 percent x 25 points   = 13 points

     Total   56 points

ASSIGNED CREDIT FOR FIRE HYDRANTS

The assigned credits earned for hydrant size, type, and installation are determined by application of the following formula:

$$CH = PH/100 \times 2$$

Where:

CH = Credit for fire hydrants

PH = The summary of points for fire hydrants

The actual assignment of credit for the community under consideration is as follows:

$$CH = 56/100 \times 2 = 1.12 \text{ percent out of a possible 2 percent for this item evaluation}$$

INSPECTION AND CONDITION OF FIRE HYDRANTS

The *Grading Schedule* indicates that the inspection and condition of fire hydrants should be in accordance with the AWWA's manual–M-17, *Installation, Field Testing and Maintenance of Fire Hydrants*, 4th Ed., 2006.

This item considers both the frequency of the fire hydrant inspections and the operating condition of fire hydrants. To obtain the maximum credit of 3 percent for this item, a community needs to inspect **all** fire hydrants twice a year and maintain all fire hydrants so there are no leaks and each fire hydrant opens easily. All fire hydrants should be visible from the street and well-located for use by a fire department pumper. Records need to be maintained for **each** fire hydrant inspection. The point credits and the factors used for this item are given below.

INSPECTION

The frequency of inspection is the average time interval between the three most recent inspections.

Frequency of Inspections	Points
Every 6 months	100
One time per year	80
Every 2 years	65
Every 3 years	55
Every 4 years	45
5 years or more	40

The points for inspection frequency are to be reduced by 10 if hydrants are not subjected to full system pressure during the inspection. Not all fire hydrants are examined during a community survey by an ISO field representative. The fire hydrants used for flow tests are evaluated carefully. Additional fire hydrants may be spot-checked during the survey of the entire community. The factors are prorated from the actual number of fire hydrants examined during the survey as outlined below:

Condition	Factor
Standard (no leaks, opens easily, conspicuous, well-located for use by a fire department pumper)	1.0
Usable	0.5
Not usable	0.0

During a community survey, the ISO field representative determines that records support the claim that fire hydrants are inspected and pressure-tested on a yearly frequency. Therefore, 80 points are earned for fire hydrant inspections.

A total of 42 fire hydrants were operated during the survey. It was determined that 35 hydrants met standard operating conditions, 5 were usable, and 2 were not usable.

This information is prorated to provide a composite factor as follows:

Standard:	35/42x1.0=0.83
Usable:	5/42x0.5=0.06
Not usable:	2/42x0.0=0.05
	_____
Total	0.94

Administrative Perspective For Community leaders

Municipal administrators should realize that NFFs throughout the city represent the dominant factor in evaluating a community water supply system. NFFs are matched to available fire flows at representative fire risk site locations. Actual water tests are conducted to determine available fire flows at 20 psi residual pressure.

Generally, the only practical way to reduce a given NFF for an existing building is to install automatic sprinkler protection. Sprinklered buildings are not considered in the evaluation of cities to develop a public protection class for insurance rating purposes.

Those involved in water supply system evaluations and community planning should carefully consider the impact of new construction on NFFs. Individual NFFs should be evaluated against the measured or designed fire flow at a new construction site. An available fire flow that is less than the NFF will have a **negative** impact on future grading surveys.

Computer evaluation of water distribution systems are important for determining flow characteristics throughout the entire water supply distribution supply pipe network. This topic is the subject of Chapter 9, and should be considered carefully for the real-time analysis of distribution systems. Caution! Computer models are only as good as their calibration to the actual size and characteristics of underground piping systems. Theoretical flow characteristics based on the design of a water system should be contrasted to the actual distribution system capability, as determined from actual flow tests conducted at least semiannually. This will identify the strengths and weaknesses of the water system clearly in meeting both consumer needs and fire-flow needs. This type of analysis can identify potential problems such as completely or partially closed valves, tuberculated pipes, system leaks, and other hydraulic gradient problems.

Few cities or smaller communities can meet all of the established NFFs at representative risk sites. It is important to know that water supply to any specific risk site can be augmented by demonstrating that water can be delivered from ground-level water sources such as lakes, ponds, rivers, and streams. Concepts of both alternative water supplies and optional water supplies are laced throughout both volumes of this series.

Water supply officials, fire officials, and municipal officials should work together to ensure that fire hydrant specifications are established to meet the following criteria for all new fire hydrants, and it may be advantageous to the adequacy and reliability of a water system to replace some of the older fire hydrants that do not conform to the following minimum criteria: 1) a 6-inch branch or lateral; 2) a pumper outlet–preferably 6 inch; 3) all city fire hydrants need to operate in the same direction; and 4) all hydrant outlets should have the same thread, again, preferably National Standard Threads.

Reference:

1. *Installation, Field Testing and Maintenance of Fire Hydrants.* 4th Ed., Denver: AWWA, 2006.

CHAPTER 6: EVALUATING THE QUALITY OF MUNICIPAL WATER SUPPLY SYSTEMS

Numerous references have been made in the first volume of this two-part series on municipal water supplies to the Environmental Protection Agency (EPA) of the Federal government and/or to individual State public health regulations on the quality or **potability** of water for drinking and cooking and any other means by which water is consumed. The regulations concerning the treatment of water have increased dramatically over the past 20 years due to the identification of new water impurities from ground water runoff and the depth from which water is taken from well sites. The EPA regulations for 2005 are presented in Volume I, Chapter 7. This chapter focuses on evaluating the quality of municipal water supplies in relation to a number of special topics presented below.

STATE AND FEDERAL REGULATIONS PERTAINING TO THE EVALUATION OF WATER QUALITY

It is very important that municipal officials associated with municipal water supplies, water supply superintendents, ranking fire officers, and water supply operators understand the basic reasons for having regulations, how they are administered to achieve water quality standards, and why compliance is essential to the well-being of all water supply consumers.

Federal Regulation

Although the regulations required by the Safe Drinking Water Act (SDWA) are of prime interest in the operation and administration of water distribution systems, municipal water supplies must adhere to the regulations of several Federal environmental and safety acts as identified below:

1. SDWA requirements.

   Requirements under the SDWA are quite extensive. The following information focuses on requirements that affect the operation of water distribution systems.

Prior to 1975, the review of public water supplies was done by each State, usually by the State health department. The SDWA was passed by Congress in 1975 for a combination of reasons. One of the primary purposes was to create uniform national standards for drinking water quality to ensure that every public water supply in the country would meet minimum health standards. Another was that scientists and public health officials had recently discovered many previously unrecognized disease organisms and chemicals that could contaminate drinking water and might pose a health threat to the public. It was considered beyond the capability of the individual States to deal with these problems.

The SDWA delegates responsibility for administering the provisions of the act to the EPA. This agency is headquartered in Washington, DC, and has 10 regional offices in major cities of the United States. Some principal duties of the agency are to:

- set maximum **allowable** concentrations for contaminants that might present a health threat in drinking water; these are called maximum contamination levels (MCLs);
- delegate primary enforcement responsibility for local administration of the requirements to State agencies;
- provide grant funds to the States to assist them in operating the greatly expanded program mandated by the Federal requirements;
- monitor State activities to ensure that all water systems are meeting Federal requirements; and
- provide continued research on drinking water contamination and improvement of treatment methods.

2. State primacy.

The intent of the SDWA is for each State to accept primary enforcement responsibility (primacy) for the operation of the State's drinking water program. Under the provisions of the delegation, the State must establish requirements for public water systems that are least as stringent as those of the EPA. The Governor has designated the primacy agency in each State. In some States the primacy agency is the State health department, and in others it is the State environmental protection agency, department of natural resources, or pollution control agency.

3. Classes of public water systems.

The basic definition of a public water system in the SDWA is, in essence, a system that supplies piped water for human consumption and that has at least 15 service connections, or serves 25 or more persons for 60 or more days of the year. Examples of water system that would **not** fall under the Federal definition are private homes, groups of fewer than 15 homes using the same well, and summer camps that operate for fewer then 60 days per year. These systems are, however, generally under some degree of supervision by a local, area, or State health department.

Figure 6-1

Classification of Public Water Systems

Source: Drinking Water Handbook for Public Officials (2003)

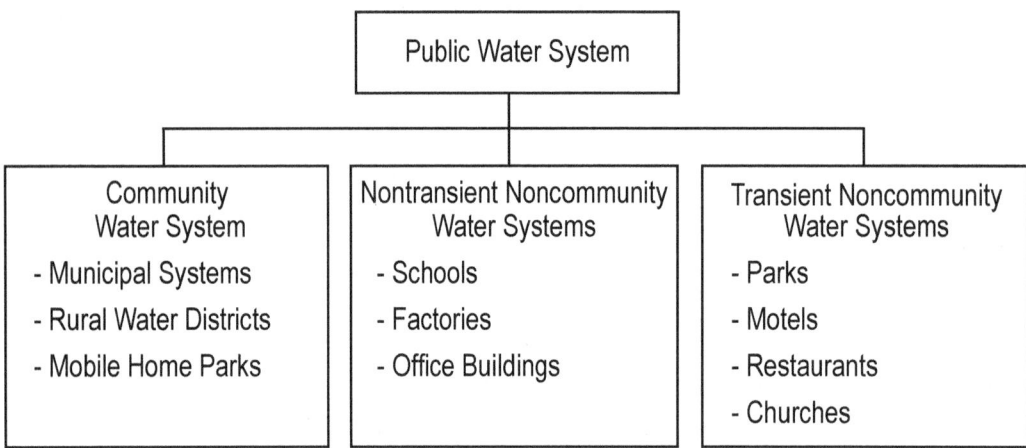

EPA has further divided public water systems into three classifications, as depicted in **Figure 6-1**:

💧 Community public water system that serve 15 or more homes. Besides municipal water utilities, this classification also covers mobile home parks and small homeowner associations that have their own water supply and serve more than 15 homes.

💧 Nontransient, noncommunity water systems or establishments that have their own private water systems, serving an average of at least 25 persons who do not live at the location, but the same people use the water for more than 6 months per year. Examples include schools and factories.

💧 Transient, noncommunity public water systems are establishments such as parks and motels that have their own water systems and serve an average of at least 25 persons per day, but these persons use water only occasionally and for short periods of time.

The monitoring requirements for community and nontransient, noncommunity systems include all contaminants that are considered a public health threat. Transient, noncommunity systems are only required to monitor for nitrate, nitrite, and microbiological contamination.

1. Regulation of contaminants.

    The National Secondary Drinking Water Regulations (NSDWRs) basically apply specify MCLs or a treatment technique requirement for contaminants that may be found in drinking water and could have an adverse health effect on humans. Specific concentration limits for the chemicals are listed, and all community and nontransient, noncommunity systems must test for their presence. If a water system is found to have concentrations of chemicals above the MCL, the system must either change its water source or treat the water to reduce the chemical concentration. Primary regulations are mandatory, and must be complied with by all water systems to which they apply.

The NSDWRs basically apply to drinking water contaminants that may affect the aesthetic qualities of the water adversely, such as taste, odor, or color. These qualities have no known adverse health effect, but they seriously affect public acceptance of the water. Secondary regulations are not mandatory, but are strongly urged by the EPA. Some State regulatory agencies have made some of the secondary limits mandatory in their States.

2. Public notification.

The SDWA mandates that the public be kept informed of noncompliance with Federal requirements by requiring that noncomplying systems provide public notification (PN). If public water systems violate any of the operating, monitoring, or reporting requirements, or if the water **quality** exceeds an MCL, the system must inform the public of the problems. Even though the problem may already have been corrected, an explanation must be provided in the news media describing the public health significance of the violation.

The language and methods of providing PN are mandated by the EPA to make sure the public is fully informed. If a system is required to provide PN, the State primacy agency will provide full instruction.

Water distribution operators should understand that, although PN is intended to keep the public informed, if it is caused by a simple mistake such as forgetting to send in the monthly samples, even on an infrequent basis, it can cause some embarrassment for the system staff. Two pieces of advice can be provided: First, pay careful attention to **individual State requirements**. If there are any problems in meeting any requirements, discuss it with the State agency representative. For instance, if the person who normally sends in the water samples is sick or on vacation, do not forget to have someone else assigned to do this work.

The second consideration is simply that, if a community is required to provide public notification, make it as positive as possible. Although a certain amount of wording is mandatory, other wording can be added to keep it from sounding overly negative.

REQUIREMENTS AFFECTING DISTRIBUTION SYSTEM OPERATIONS

Some of the significant Federal requirements that particularly apply to water distribution system operations may be summarized under the following subtopics:

Monitoring and Reporting

To ensure that drinking water supplied by **all** public water supply systems as defined by the EPA meet Federal and State requirements, water system operators are required to collect samples regularly and have the water tested. The regulations specify minimum sampling frequencies, sampling locations, testing procedures, methods of keeping records, and frequency of reporting to the State. The regulations also mandate special reporting procedures to be followed if a contaminant exceeds an MCL.

All systems must provide periodic monitoring for microbiological contaminants and some chemical contaminants. The frequency of sampling and the chemicals that must be tested for depend on the physical size of the water system, the water source, and the history of analyses. General sampling procedures are covered in more detail under the topic of Public Health Considerations to follow.

State policies vary on providing laboratory services. Some States have laboratory facilities available to perform all required analyses or, in some cases, a certain number of the required analyses for a system. In most States, there is a charge for all or some of the laboratory services. Sample analyses that are required and cannot be performed by a State laboratory must be taken or sent to a State-certified private laboratory.

If the analysis of a sample exceeds an MCL, resampling is required, and the State should be contacted immediately for special instructions. There is always the possibility that such a sample was caused by a sampling or laboratory error, but it must be handled as though it actually was caused by contamination of the water supply.

The results of all water analyses must be periodically sent to the State of origin. Failure to have the required analysis performed or to report the results to the State usually will result in the water system being required to provide PN. States typically have special forms for submitting data, and specify a number of days following the end of the monitoring period by which the form is due. Lab report requirements are listed in **Figure 6-2**. State regulators also may require other information for their own records and documentation.

Figure 6-2

Lab Report Summary Requirements

Type of Information	Summary Requirement

Sampling information	Date, place, and time of sampling
	Name of sample collector
	Identification of sample
	  • Routine or check sample
	  • Raw or treated water

Analysis information	Date of analysis Laboratory conducting analysis Name of person responsible for analysis Analytical method used Analysis results

There are also specific requirements for the length of time a water system must retain records. **Figure 6-3** lists the recordkeeping requirement mandated by the EPA.

Figure 6-3

Recordkeeping Requirements

Type of Records	Time Period
Bacteriological and turbidity analysis	5 years
Chemical analysis	10 years
Actions taken to correct violations	3 years
Sanitary survey reports	10 years
Exemptions	5 years following expiration

The Lead and Copper Rule

The Lead and Copper Rule (LCR) is a requirement that applies specifically to water distribution systems. The rule was created because it was found that, even though water entering a distribution system may be of acceptable quality, the quality may change by the time it gets to the consumer's tap. If the water is aggressive, lead and copper may dissolve from water piping systems. The requirements were prompted by findings that showed consumption of even minute quantities of lead may have an adverse health effect on all humans, and especially on children.

The rules require that samples be collected at customer's taps after the water has remained undisturbed in the water piping for at least 6 hours. If the samples are found to contain excessive levels of lead or copper, the water system must implement corrosion control techniques. Specific monitoring procedures to be used and monitoring frequency are provided to each system by the State primacy agency.

General Disinfection Requirements

Disinfection is absolutely required for all water systems using surface water sources. Various chemicals other than chlorine can be used for treatment of surface water, but as the water enters the distribution system, it must carry a continuous chlorine residual that will be retained throughout the distribution system. Water samples from points on the distribution system must be analyzed periodically to make sure an adequate chlorine residual is being maintained.

In spite of the fact that use of chlorine has almost completely eliminated occurrences of waterborne diseases in the United States, there is no concern for byproducts formed when chlorine reacts with naturally occurring substances in raw water (such as decaying vegetation containing humic and fulvic acids). The first group of byproduct chemicals identified was tri-halo-methane (THM), a group of organic chemicals that are known carcinogens (cancer-forming) to some animals, so they are assumed also to be carcinogenic to humans. Other byproducts of disinfection have been identified that may be harmful, and there also is concern now that disinfectants themselves may cause some adverse health reactions.

Water treatment systems have been trying for several years to limit formation of THM without impairing the bacteriological safety of water. Proposed new regulation of disinfectants is called the Disinfectants-Disinfection By-Product Rule (D-DBP Rule). Methods of how best to disinfect water to protect the public are still under study, and will result in new chemical addition and monitoring requirements in the future.

Disinfection of Groundwater

Federal regulations do not currently (2005) require disinfection of ground water unless the well has been designated by the individual State as vulnerable to contamination by surface water (i.e., ground water under the direct influence of surface water). These are generally shallow wells. Many States, though, have been phasing in their own individual requirements for disinfection of various sizes, types, or classes of well systems.

The SDWA amendments of 1996 specifically state that the EPA must promulgate regulations requiring disinfection as a treatment technique **as necessary for groundwater systems**. The final rules on this matter probably will give the individual States authority to allow well-water systems that are properly constructed and operated, and meet other criteria, to forgo applying disinfection treatment.

Consumer Confidence Reports

One of the very significant provisions of the 1996 SDWA amendments is the consumer confidence report (CCR) requirement. The purpose of the CCR is to provide all water customers with basic facts regarding their drinking water so that individuals can make decisions about water consumption based on their personal health. This directive has been likened to the requirement that packaged food companies disclose what is in their food product.

The reports must be prepared yearly by every community water supply system. Water systems serving more than 10,000 people must mail the report to customers. Small systems must notify customers as directed by the State primacy agency. Beginning in the year 2000, reports were to be delivered by July 1 of the calendar year.

A water system that only distributes purchased waster (i.e., a satellite system) must prepare the report for their consumers. Information on the source water and chemical analyses must be furnished to the satellite system by the system selling the water (parent company).

Some States are preparing much of the information for their water systems, but the system operator still must add local information. Templates for preparing a report also are available from the American Water Works Association (AWWA) and the National Rural Water Association (NRWA). Water system operators should keep in mind that CCRs provide an opportunity to educate consumers about the sources and quality of their drinking water. Educated consumers are more likely to help protect drinking water sources and be more understanding of the need to upgrade the water system to make their drinking water safe.

Environmental Protection Agency Regulation Information

Current information on EPA regulations can be obtained by contacting the Safe Drinking Water Hotline at 800-426-4791. Also, see the Office of Ground Water and Drinking Water Web page at: http://www.epa. gov/safewater/standards.html

State Regulations

Under the provisions of primacy delegation, each State must have requirements applying to public water systems that are at least as stringent as those set by the EPA. States occasionally establish requirements that are more stringent. Federal requirements are only for factors that the EPA considers directly related to public health. So, in addition to the Federal requirements, each State also establishes other requirements to ensure proper water system operation.

Operator Certification

One requirement of the 1996 SDWA amendments is that the EPA must establish minimum standards for State operator certification programs. Most States have had some form of certification for water system operators but, unfortunately, each State has its own idea of how operators should be classified so there has been little national consistency.

The new requirements will not correct the inconsistency, but will require most States to make some changes in their certification programs. Among the more important requirements are that each water supply system must at **all** times be under the direct supervision of a certified operator. Each operator must have a high school or equivalent education and pass an examination to receive certification, and the specific State must establish training requirements for certification renewal. Most States have a separate certification class for distribution system operators.

Cross-Connection Control

The individual States also generally promote cross-connection control programs for all water systems. Many States have their own cross-connection control manuals and assist water systems in setting up local programs.

Construction Approval

The SDWA requires States to review plans for water system construction and improvements. In general, plans and specifications for the proposed work must be prepared by a professional engineer (PE) and submitted for approval before work begins. State engineers review the plans for suitability of materials, conformance with State regulations, and other factors.

Some States allow small distribution system additions without approval, or allow approval after construction. State regulations should be reviewed to ensure compliance with requirements.

Sanitary Surveys

A sanitary survey is an onsite inspection of water system's facilities and operation. The survey is usually performed by a State employee, but the State also may contract with another person to do the work. Survey visits range in frequency from yearly to once every several years, depending on the water source and treatment process being used, size of the distribution system, history of compliance with monitoring and reporting requirements, and various other factors.

A sanitary survey usually involves a review of operating methods and records and a physical review of various facilities and equipment. The survey is designed to note problems or deficiencies that could cause contamination of the water supply or interrupt continuity of service. Surveys also produce recommendations on needed programs and changes to improve water quality, quantity, and reliability. A summary of the observations and suggestions or directives resulting from the survey usually is furnished in writing to the water system owner or person in charge.

Technical Assistance

One of the staff functions of the State drinking water program is to provide technical assistance to water system operators. Field staff with training and experience usually are available to provide advice and assistance. If possible, they will provide advice over the phone, but if the problem is of sufficient magnitude, they will arrange personal visits. They may also, on some occasions, suggest other sources of information or assistance.

Enforcement

Because of the direct relationship between drinking water quality and public health, it is rare for anyone to purposely disregard State and Federal regulations. Most violations of regulations are due to not understanding requirements or forgetting something that must be done.

The SDWA requires States to use enforcement actions when Federal requirements are violated. If the State does not take appropriate action, the EPA is prepared to step in and do it. Minor infractions are handled by public notification, but intentional disregard for requirements can result in substantial monetary fines.

PUBLIC HEALTH CONSIDERATIONS

Sources of Contamination

Water may be contaminated by biological, chemical, or radiological agents. Although water is supposed to be free of harmful and objectionable contamination when it enters the distribution system, there are a number of opportunities for water quality to change before it reaches the customer's taps. It is the distribution system operator's responsibility to ensure safe and pleasing water is delivered to each tap.

Biological Contamination

Pathogenic organism is the term often used to cover all organisms that may cause human sickness or death. All pathogenic organisms have their origin in fecal waste from humans or other warm-blooded animals. Although most disease-causing organisms die quickly after being released to the environment, there are a few that can remain viable for days or even weeks.

Some of the historic waterborne diseases were typhoid, cholera, and dysentery. These are now rarely found in the United States, but the organisms are now more likely to cause contamination in water are Legionella, Cryptosporidium cysts, and Giardia lamblia cysts.

1. Bacteria contamination.

    Although bacteria are the most plentiful of all living organisms, only a few are pathogenic. Many bacteria are helpful and even necessary for everyday living. Disease-causing bacteria reside in the intestinal tracts of humans and warm-blooded animals. It is almost impossible to specifically identify the disease bacteria, so water samples are tested for the presence of any bacteria of the coliform group, which is used as an indicator of fecal contamination. The theory is, if any coliform bacteria are present, there could be disease-causing bacteria among them.

2. Viral contamination.

    Viruses that are of particular concern as causes of waterborne diseases include infectious hepatitis, polio, and several types that cause gastrointestinal disease. Tests for specific viruses are very difficult, but because viruses originate from the fecal matter of warm-blooded animals, the lack of coliform bacteria in a sample is also taken as an indication that there are no harmful viruses present.

Radiological Contamination

There is some slight chance of water being contaminated by radioactive industrial wastes that have been improperly disposed of, but government regulations on handling and disposal of radioactive materials are very strict, so this danger is generally minimal.

There, are, though, many ground water aquifers in the United States that have naturally occurring radium. The EPA has set limits on the concentration of radium that is considered a danger to public health. Water systems with wells drawing water from these aquifers must either change their water sources or install treatment to reduce the radium level of water furnished to customers.

Many wells also produce water containing quantities of radon gas. This is considered a danger to health because the radon is released into homes as the water is used in showers and other applications, and continued inhalation of radon gas is considered to contribute to the risk of lung cancer. The EPA has established limits for the concentration of several organic chemicals in drinking water, so water systems with well water exceeding the limits will have to install treatment for radon removal.

Chemical Contamination

Chemical contamination of drinking water can be due either to naturally occurring chemicals or to wastes from human activities. Limits on chemicals such as arsenic, barium, and cadmium have been established in the Primary Drinking Water Regulations. It has also been found in recent years that some manufactured organic chemicals disposed of on the ground over the years do not disintegrate as previously supposed.

Instead, some of these chemicals have been found to travel considerable distances in aquifers and are now found in wells at concentrations considered a danger to public health.

The EPA has established limits for the concentration of several organic chemicals in drinking water, and limits for additional chemicals probably will be added in the future as research reveals that they pose a threat to public health.

WATER QUALITY MONITORING

Although most water quality monitoring is related to ensuring proper quality of the source water or treatment processes, many of the samples are collected from the distribution system. Thus, sample collection often becomes a duty of distribution system personnel. The reason for collecting samples from the distribution system is that there are some opportunities for water quality to change after it enters the distribution system and, under the requirements of the SDWA, it is the duty of the water purveyor to deliver water of proper quality to the consumer's tap.

Methods of Collecting Samples

There are two basic methods of collecting samples: 1) grab sampling and 2) composite sampling. A **grab sample** is a single volume of water collected at one time from a single place. To sample water in the distribution system, a faucet is used to fill a bottle. The sample represents the quality of the water **only** at the time the sample was collected. If the quality of the water is relatively uniform, the sample will be quite representative. If the quality varies, the sample may not be representative.

A **composite sample** consists of a series of grab samples collected from the same point at different times and mixed together. The composite then is analyzed to obtain the average value. If the composite is made up of equal-volume samples collected at regular intervals, it is called a **time-composite** sample. Another method is to collect samples at regular time intervals, but the size of each grab sample is proportional to the flow at the time of sampling. This is called a **flow-proportional composite** sample.

Although composite sampling appears to be a good idea because it provides an average of water quality, it cannot be used for more analyses of drinking water quality because most parameters are not stable over a period of time.

Sample Storage and Shipment

Care always must be taken to use the exact sample containers specified or provided by the laboratory that will be doing the analysis. Most sample containers are now plastic to avoid the possibility of glass breaking during shipment. There are some samples for organic chemical analysis that must be collected in special glass containers because some of the chemical might be lost by permeating through the walls of plastic containers.

Sample holding time before analysis is quite critical for some parameters. If a laboratory receives a sample that has passed the specified holding time, it is supposed to declare the sample invalid and request resampling. Some samples can be refrigerated or treated once they arrive at the laboratory to extend the holding time, allowing the laboratory a few more days before analysis must be completed.

Many laboratories do not work on weekends, so this should be taken into consideration when sending samples. Bacteriological analysis must, for example, be performed immediately by the laboratory. The best time to collect and send these samples is on a Monday or Tuesday so they will reach the laboratory by midweek. Samples should be sent to the laboratory by the fastest means available, such as Express Mail, Fed-X, or Parcel-X.

Sample Point Selection

Samples are collected from various points in the distribution system to determine the quality of water delivered to consumers. In some cases, distribution system samples may be significantly different from samples collected as the water enters the system. For example, corrosion in pipelines, bacterial growth, and algae growth in the pipes can cause increases in color, odor, turbidity, and chemical content (e.g., lead and copper). More seriously, a cross-connection between the distribution system and a source of contamination can result in chemical or biological contamination of the water.

Most of the samples collected from the distribution system will be used to test for coliform bacteria and chlorine residual. The two primary considerations in determining the number and location of sampling points are that they be:

1. Representative of each different source of water entering the system (i.e., If there are several wells that pump directly into the system, samples should be obtained that are representative of the water from each one).

2. Representative of the various conditions within the system (such as dead ends, loops, storage facilities, and each pressure zone).

The required number of samples that must be collected and the frequency of sampling depend on the number of customers serviced, the water source, and other factors. **Specific** sampling instructions **must** be obtained from the State primacy agency.

Sample Faucets

Once representative sample points have been identified on the water distribution system, specific locations having suitable faucets for sampling must be identified. If suitably located, public buildings and the homes of utility workers and government officials are convenient places to collect samples. Arrangement should be make to collect 10 percent of the samples from local businesses, and request permission to gather samples near the final domestic tap on dead-end mains.

The following is a list of types of sampling faucets that **should not** be used:

- any faucet located close to the bottom of a sink, because containers may touch the faucet;
- any leaking faucet with water running out from around the handle and down the outside;
- any faucets with threads, **such as a sill cock**, because water generally does not flow smoothly from them and may drip contamination from the threads;
- any faucet connected to a home water-treatment unit, such as a water softener or carbon filter; and
- drinking fountains.

It also is best to try to find a faucet without an aerator. If faucets with aerators must be used, follow the State recommendation on whether or not the aerator should be removed for sampling. Some years ago, it was recommended that faucets be flamed before samples were taken. This generally consisted of running the flame from a propane torch over the outside of the faucet to kill any germs that may be present. Problems with the process included customers objecting because of possible damage to the finish on their faucets and, in many cases, faucets that look like metal are actually plastic, creating a fire hazard. It has been determined that, if proper faucet and technique sampling are used, flaming is not desirable or recommended.

Each sample point must be described in detail on the sample report form, not just the street address, but which faucet, in which room. If resampling is necessary, the same faucet used for the first sample must be used.

When it is necessary to establish a sampling point at a location on the water system where no public building or home gives access for regular sampling, a permanent sampling station can be installed as depicted in **Figure 6-4**.

Figure 6-4

Example of a Permanent Sampling Station

Courtesy of Gil Industries, Inc.

Sampling Collection

For collection of bacteriological and most other samples, the procedure is to open the faucet so that it will produce a steady, moderate flow as depicted in **Figure 6-5**. Opening the faucet to full flow for flushing is not usually desirable because the flow may not be smooth, and water will splash up onto the outside of the spout. If a steady flow cannot be obtained, that specific faucet should be not be used.

Figure 6-5

Sample Faucet set to Produce a Steady, Moderate Flow

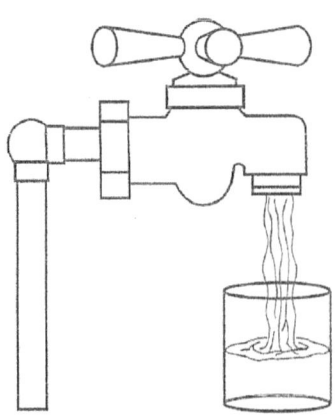

The water should be allowed to run long enough to flush any stagnant water from the house plumbing, which usually takes 2 to 5 minutes. The line usually is clear when the water temperature drops and stabilizes. The sample then is collected without changing the flow setting. The sample container lid should be held (not set down on the counter or ledge) with the threads down during the sample collection and replaced immediately. The sample container then should be labeled.

The exception to the above procedure is sampling for lead and copper analysis. These are first-draw samples, and require special procedures.

Bottles to be used for collection of bacteriological samples should not be rinsed before they are filled. These bottles usually are prepared with a small quantity of sulfate at the bottom to immediately stop the action of the residual chlorine in the water.

Special Purpose Samples

It is occasionally necessary to collect special samples, particularly in response to customer complaints, such as taste and odor issues. To check on this type of complaint, one sample should be collected immediately as the tap is opened to be representative of water that has been in the plumbing system, then a second sample should be collected after the line has been flushed. It sometimes is helpful to collect both hot and cold water samples in this manner. These samples can be used to identify whether the problem is in the customer's plumbing system or coming from the water distribution system. Many customers' complaints of taste, odor, or color are found to be from their own water heaters, water softeners, or home water-treatment devices.

DISTRIBUTION SYSTEM WATER QUALITY PROBLEMS

Turbidity

Turbidity is caused by particles suspended in water. These particles scatter or reflect light rays, making the water appear cloudy. Turbidity is expressed in nephelometric turbidity units (ntu) and a reading in excess of 5 ntu is generally noticeable to water system customers.

Besides the appearance being unpleasant to customers, turbidity in water is significant from a public health standpoint because suspended particles could shelter micro-organisms from the disinfectant and allow them to still be viable when they reach the customer. EPA regulations direct that, for most water systems, the turbidity of water entering the distribution system must be equal or less than 0.5 ntu in at least 95 percent of the measurements taken each month. At no time may the turbidity exceed 5 ntu.

Turbidity changes in the distribution system can indicate developing problems. Increases in turbidity may be caused by changes in velocity or inadequate flushing following main replacement.

Hardness

Hardness is a measure of the concentration of calcium and magnesium in water. Water hardness usually comes from water contacting rock formations, such as water from wells in limestone formations. Soft ground water may occur where topsoil is thin and limestone formations are sparse or absent. Most surface water is of medium hardness.

Hard and soft water are both satisfactory for human consumption, but customers may object to very hard water because of the scale it forms in plumbing fixtures and on cooking utensils. Hardness is also a problem for some industrial and commercial users because of scale buildup in boilers and other equipment.

Water generally is considered most satisfactory for household use when the hardness is between 75 and 100 mg/L as calcium carbonate ($CaCO_3$). Water with 300 mg/L of hardness usually is considered **hard**. Very soft water of 30 mg/L or less is found in some section of the United States. Soft water usually is quite corrosive, and may have to be treated to reduce the corrosivity.

Iron

Iron occurs naturally in rocks and soils and is one of the most abundant elements. It occurs in two forms. Ferrous iron (Fe_2) is in a dissolved state, and water containing ferrous iron is colorless. Ferric iron (Fe_3) has been oxidized, and water containing it is rust-colored. Water from some well sources contains significant levels of dissolved iron, which is colorless, but rapidly turns brown as air reaches the water and oxidizes the iron.

There are no known harmful effects to humans from drinking water containing iron, but NSDWR suggest a limit of 0.5 mg/L. At high levels, the staining of plumbing fixtures and clothing becomes objectionable. Iron also provides nutrient source for some bacteria that grow in distribution systems and wells. Iron bacteria, such as Gallionella, cause red water, tastes and odors, clogged pipes, and pump failure.

Whenever tests on water samples show increased iron concentrations between the point where water enters the distribution system and the consumer's tap, either corrosion, iron bacteria, or both are probably taking place. If the problem is caused by bacteria, flushing mains, shock chlorination, and carrying increased residual chlorine are alternatives to consider.

Manganese

Manganese in ground water creates problems similar to iron. It does not usually discolor the water, but will stain washed clothes and plumbing fixtures black; this is very unpopular with customers. Consumption of manganese has no known harmful effects on humans, but the NSDWR recommend a concentration not to exceed 0.05 mg/L to avoid customer complaints.

WATER QUALITY SAFEGUARDS

The **critical** safeguard for water distribution system operations are

- continuous positive pressure in the mains; 20 pounds per square inch (psi) minimum residual pressure is recommended;
- maintenance of chlorine residual;
- cross-connection control; and
- frequent testing.

Continuous positive pressure as recommended above is absolutely necessary to prevent back siphonage and the entry of contaminants into the water system. This can be achieved primarily by maintaining an adequate water supply and storage capable of meeting peak water demands. If water demands are so great during peak demand periods that pressure declines in parts of the systems, either water use must be restricted or the water system must be upgraded to be capable of supplying more water.

System pressure also may be reduced during a main break because of the large amount of escaping water. The best safeguards against having serious pressure loss during a main break are to have adequate system storage and to be well-organized to shut down the leaking section of water main swiftly. The later involves having personnel on call at all time to respond to emergencies, knowing where all the valves are, and having a valve exercise program so that valves are sure to operate when needed.

The ultimate proof of the bacteriological safety of the water in the distribution system comes through frequent sampling. Samples collected to meet State requirements should be considered a minimum. Additional samples should be collected following construction and repair work as well as in response to customer complaints that could be the result of water system contamination. A distribution system can become contaminated from an outside source by accident or intention in the framework of the world climate today. Contamination problems need to be identified and appropriate action taken immediately after detection.

Chapter 7: Impacts of Fire Flow on Distribution System Water Quality, Design, and Operation

Background Statement

In 2002, The American Water Works Association (AWWA) Research Foundation and KIWA of the Netherlands convened a panel of engineers, scientists, university professors, municipal waterworks officials, senior fire officers, and insurance industry personnel to study, explore, and evaluate, as this Chapter title indicates, *Impacts of Fire Flow on Distribution Systems Water Quality, Design, and Operations*. This study was funded jointly by the AWWA, the AWWA Research Foundation, and KIWA of the Netherlands who assume no responsibility for the content of the research study reported in a publication by the Chapter title above or the facts and opinions of the panel participants as expressed in the report.

The following information is excerpted from this study because the information presented is essential to understanding current issues associated with public-sector water supplies and the potential impact that the private sector has on needed water supplies for fire protection. It should be understood that there are two approaches to the confinement, control, and extinguishment of developing fires in structures. The conventional approach is to have a public fire department respond to a structure fire with engine companies and ladder companies to handle everything from a wastebasket fire to a fully involved structure fire. The alternative approach is to install automatic sprinkler systems in both residential and commercial properties either to extinguish a developing fire immediately, or to confine a fire to the room of origin to be completely extinguished by a fire department.

The installation of automatic sprinkler systems transfers a significant portion of the cost of water supplies for fire protection to the private sector. The private sector assumes the cost of the installation of automatic sprinkler systems, maintenance, and periodic testing. There are economic incentives in the form of insurance premium reductions for commercial property owners with installed and maintained sprinkler systems, and even some insurance incentives for residential sprinkler system installations.

As stated in previous chapters, the primary objective of a public water system is to provide sufficient potable water at an acceptable level of water quality, now determined by the Environmental Protection Agency (EPA) in the United States. The decision for a public water supply to provide fire flows can have significant impacts on the design and operation of the systems. This is why a large number of small villages and towns under approximately 5,000 population do not have fire hydrants installed on a small water distribution system.

Typically, electing to provide fire flows and fire hydrants results in increased water supply pipe diameters, leading to higher capital costs and greater provision for reliability and redundancy in the distribution system. It may also, however, have some negative water quality implications. This oversizing to meet what some consider to be relatively infrequent fire events can result in increased water resident times in larger size pipe, thus increasing the possibility of residual disinfectant loss, and enhancing the formation of disinfection byproducts and bacterial growth in the water mains. Larger diameter pipes also result in lower water flow velocities in the water system that lead, in turn, to the deposit of sediments.

Background

The study group determined that, in the United States, there is no legal requirement for the water distribution system to provide fire flows. Municipalities and townships may develop ordinances that require water systems to provide fire protection. At the same time, an inadequate fire protection system provides a false sense of security. Fire-flow requirements generally are based upon fire codes developed by independent groups including the Building Officials and Code Administrators International, Inc. (BOCA), the International Conference of Building Officials (ICBO), the Southern Building Code Congress International, Inc. (SBCCI), and the National Board of Fire Underwriters up to 1970. Since 1970, the insurance criteria have been developed by the Insurance Services Office, Inc. (ISO).

In the absence of a fire, the obligation to provide fire protection substantially extends the time that the water, after treatment, resides in the distribution system, including its service reservoirs, before it reaches the users. Degradation of water quality in conventional potable water distribution systems has been shown to be a function of the length of time that water is retained in the distribution system and the very slow velocities of water in the lines that feed residential areas, to the point where they protect microorganisms in the water and, over time, restrict flow capacities of the pipes.

Water quality changes within the distribution system are of increasing concern. Water distribution system design often is dictated by the need to provide fire flows. This leads to oversizing of the system for most normal conditions. An oversized system increases the resident time of water, increasing the possibility for depletion of disinfectant residual and the formation of disinfection byproducts. An oversized water distribution system also increases the associated capital and operation and maintenance (O&M) costs. This is particularly true in the case of pumping, when elevated tanks are maintained at high levels in order to provide sufficient fire-flow capacity and minimum pressures on fire hydrants. Therefore, there is a clear need to evaluate fire-flow requirements and to determine improvements in water quality and reductions in capital and O&M costs that could be achieved if fire-flow requirements are reduced or eliminated.

STUDY OBJECTIVES AND SCOPE

The objectives of the stated project were as follows:

- determine the economic and water quality impacts of designing systems to meet fire-flow requirements;
- review national and international criteria for providing fire flows; and
- identify alternative methods and technologies for firefighting appropriate for present-day situations.

This review of the study report addresses each of these topics in order.

The Impact On Water Quality by Designing Water Distribution Systems to Meet Fire-Flow Requirements

In order to evaluate the economic and water quality impacts associated with fire-flow requirements, hydraulic water quality modeling of hypothetical and actual distribution systems was conducted. The EPA's EPANET model was used to conduct the modeling. The hydraulic evaluation considered changes in infrastructure sizing and operating practices, primarily related to storage volumes, which could be realized if fire-flow requirements were reduced or eliminated. The economic impact of these changes was estimated to evaluate the financial impacts of providing fire flows. The key water quality parameter considered was water age. The residence time of water in the distribution system can be a key indicator of water quality, as it plays a major role in determining disinfectant residuals and disinfection byproduct formation. The significant findings follow.

Mathematical models were applied to various systems designs and different levels of fire flow in order to study the resulting economic and water quality impacts. The following overall approach was used in conducting this analysis:

1. A modified version of EPANET distribution system model (Rossman, 2000) was applied in steady-state mode to a specified network design to determine if it would meet minimum pressure requirements under: a) maximum day demand plus fire-flow requirements, and b) peak hour demand conditions.

2. Three levels of fire protection were studied.

   a. The first level is maximum day demand plus fire-flow capacity in which fire flows meet the Needed Fire Flow (NFF) requirements as defined by the ISO for public protection requirements. (See Chapter 5 for details on the ISO NFF calculations.)

   b. The second level is maximum day demand plus reduced fire-flow capacity associated with the hose stream requirements that are needed to augment fire protection provided by automatic sprinkler systems.

   c. The third level is no fire-flow requirement (i.e., peak hour demand for consumer consumption). This alternative would be consistent with a dual water system, briefly discussed under alternative technologies later in this review, and in Chapter 8 of this manual.

3. A modified EPANET model was applied using an interactive approach in which pipe diameters were downsized under the constraint that minimum pressure requirements were met under both peak hour conditions and maximum day plus fire-flow requirements at all demand nodes. This step was repeated for the three levels of fire protection described in the previous step.

4. The EPANET model was applied in extended period simulation (EPS) mode to determine distribution system storage requirements and to ascertain that minimum pressures and flows could be met over the full duration of a fire (i.e., 2 to 4 hours).

5. The EPANET model was applied in the EPS mode to estimate water age under minimum daily demand conditions in order to calculate the average demand-weighted water age.

6. The results of the EPANET minimum day demand application were examined to determine whether sufficiently high velocities occurred to ensure self-cleaning of the pipes. The criterion used by KIWA, that a minimum velocity (1.3 feet per second) occurs at least once a day, was employed as the evaluation measure.

7. A cost model developed by EPA (Gumerman, Burris, and Burris, 1992; Clark, et al., 2001) was applied to determine the capital costs of pipes, tanks, and other facilities for the various distribution system designs associated with each level of fire protection.

8. The redundancy and reliability of each of the distribution system designs was qualitatively evaluated.

In addition to the standard procedures outlined above, a tank water age model, CompTank, developed as part of an earlier AWWA Research Foundation sponsored project (Grayman, et al., 2000) was applied to develop general guidelines for storage requirements to meet various levels of fire protection.

EPANET Distribution System Model

Hydraulic analysis of flows and pressures in a distribution system has been a standard form of engineering analysis since its development by Dr. Hardy Cross in 1936. Water distribution system computer models have been in use since the middle 1960s and have evolved into sophisticated, user-friendly tools that are capable of simulating large distribution systems. (Walski, Chase, and Savic, 2001) In more recent years, the ability to model water utility and water age has been added to hydraulic models. (Clark and Grayman, 1998) Many commercial models offer a wide range of capabilities in distribution system modeling. EPANET is an open-structured, public domain hydraulic and water quality model developed by the EPA, and is used worldwide. (Rossman, 2000) EPANET was selected for the research project.

In order to facilitate the examination of required pipe sizes, the standard EPANET model was modified for use in the study project. This modified version allows the user to examine all nodes quickly in terms of their ability to deliver specified fire-flow quantities while meeting a specified pressure requirement at the fire-flow node; 20 pounds per square inch (psi) was selected. The model also may be used to automatically downsize pipes to the minimum diameter required to deliver a specified fire flow plus normal usage.

Costing Model

The EPA has developed a model consisting of cost equations for pumps and pump stations, including new facilities and expansion of existing pipes. (Clark et al., 2001) The model represents the base construction cost data for the purchase and installation of an item, such as a particular type of pump. Cost estimates can be developed based on a series of separate items that can be added to the base installed costs. To convert the base construction cost into capital cost, the cost data must be aggregated for the entire project, and additions made that include the following items:

- general contractor's overhead and profit;
- engineering;
- land or right-of-way acquisitions;
- legal, fiscal, and administrative costs; and
- interest during construction.

O&M requirements for pumping stations include electrical energy, maintenance materials, and labor. Total operating and maintenance cost is a composite of the energy, maintenance material, and labor costs.

Observations from Model Runs

As consumer demand increases, peak demand flow becomes a controlling factor and pipe size is determined by peak flow rather than maximum demand day plus fire-flow requirements. When peak flow is the controlling factor, then the system should also meet fire-flow requirements. As peak flow decreases, water demand decreases, a system design based on peak flow will not necessarily meet fire-flow requirements.

A network design using a looped feeder system results in overall smaller diameter pipe sizes and lower water ages than either the regular grid system or the branching network system. However, the branching network results in significantly lower costs than the other networks because of the reduction in total length of pipe, despite the larger pipe diameters required. This may explain why many distribution subdivisions are built with branching-type systems rather than looping systems. It was further identified that the nonlooped configuration used in this analysis results in longer connection lines, as compared to the looped configurations.

Examination of the branching systems shows that such systems provide no service redundancy and are thus inherently less reliable (i.e., more vulnerable to service shutdowns) than looping systems. In a branching system, any outage along a single pipe results in loss of service to all downstream consumers. In a looping system, alternative paths generally will provide some degree of service even if one pipe is out of service.

As pipe size increase to meet fire-flow requirements, the average maximum velocities in individual pipes tend to decrease, resulting in a greater potential for sedimentation, deposition, and water age. This conclusion is supported by the larger percentage of pipes that do not meet the KIWA velocity goal (minimum velocity of at least 1.3 feet per second (fps) at least once a day as the fire-flow requirement increases).

Analysis of Distribution System Storage

Storage is an important component in almost all distribution systems. It can serve the following purposes (Walski, 2000):

◆ **Equalization**: Demands in a water system generally vary over the course of the day while water utilities prefer to operate their treatment facilities at a relatively constant rate. Distribution system water tanks and reservoirs frequently operate in fill-and-draw operation over the course of the day, thus providing the storage to accommodate the constant variations in water supply and demand. However, time-of-day variations in energy pricing may influence when wells and pumps are operated.

◆ **Pressure maintenance**: The water level in tanks and reservoirs largely determine the pressure in areas served by a storage facility. In order to provide sufficient pressure in many situations, particularly at the top of the pressure zone, the water level in a storage facility must be maintained within a specific range.

◆ **Fire storage**: Required or NFFs can be provided through a combination of fire storage in water tanks and reservoirs or through larger transmission lines and increased treatment capacities. In many water systems it appears that fire storage is the more economical means of meeting fire flow requirements. Dedicated water in storage for fire protection needs to be recycled weekly to prevent excessive aging and sedimentation. This can be done by bringing the tank "on line" and refilling from the top using a total flow meter to replenish the gallons of water drained from the tank. This technique keeps the tank 90 percent full at all times.

◆ **Emergency storage**: In addition to fires, other emergencies such as power outages, equipment failures, water main breaks, and temporary loss of water supply facilities can result in insufficient water supply within the distribution system. Storage provides a mechanism for providing water under such emergency conditions.

In order to accommodate the various purposes of a water tank or reservoir, they are generally designed with a given amount of capacity targeted to each of the following specific uses:

- equalization storage: storage used to allow for normal fill and draw patterns;
- ineffective (passive) storage: storage used to provide the minimum pressure requirements;
- fire storage: storage to meet sufficient, required, or NFFs; and
- emergency storage: storage reserved for emergencies other than fires.

The actual storage requirements in each of these categories vary in different storage facilities based on local regulations, operating conditions, and hydraulic conditions. It should be understood that at some water utilities, fire storage and emergency storage maybe combined, based on the assumption that both fires and other emergencies are rare events and the simultaneous occurrence of fires and other emergencies is unlikely. With the added consideration of potential terrorist threats, the probability of simultaneous events may need to be considered in contingency planning for large city water supply systems.

The primary emphasis of the study under review is to evaluate the impact of fire-flow requirements on water storage capacity. Storage capacity is aggregated into three categories: 1) equalization, 2) fire demand, and 3) reserve storage. Reserve storage corresponds to capacity that is not used for either a joint equalization or simply fire-flow demand and includes both ineffective storage and emergency (other than fire) storage. In order to develop a general relationship, the following assumptions were made

- The tank or reservoir is completely and instantaneously mixed during the fill cycle.

- The storage facility operates with a 12-hour draw and 12-hour fill cycle at constant fill and draw rate with the water level variation over the full range of the equalization storage volume.

- The water level in the tank is at its maximum level at the start of the draw cycle and again at the end of the fill cycle.

The types of storage and the fill-and-draw pattern used in the analysis are illustrated in **Figure 7-1**.

Figure 7-1

Factors Used in the Analysis of Storage

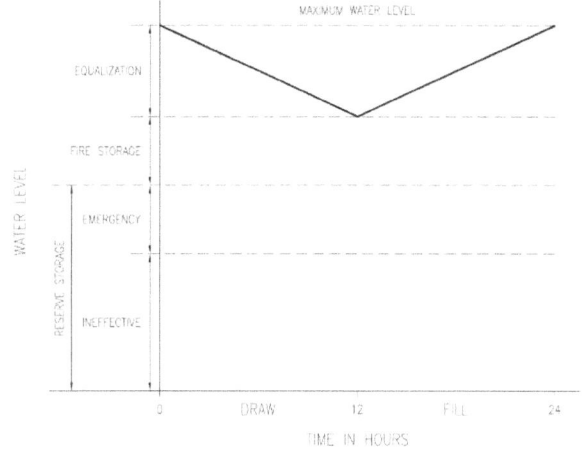

Though primarily designed for the hydraulic purposes outlined above, storage also can have a negative effect on water quality due to age, as previously identified. Loss of disinfectant residual, bacterial regrowth, taste and odor production, and formation of disinfectant byproducts are potential water quality problems associated with water aging. Improper mixing in storage facilities can exacerbate the water aging problems by creating dead or stagnant zones of even older water. (Grayman et al., 2000) A quantitative assessment of the impacts of providing fire storage on water age is provided in **Figure 7-1**.

Study Set: Summary and Conclusions

Designing a water system with sufficient capacity to meet fire-flow requirements or NFFs can result in major economic and water quality impacts. Typically these impacts are manifested in increased pipe diameters, greater provision for reliability and redundancy in system design, increased system costs, and increased potential for reduced water quality.

A methodology has been developed for approximating minimum pipe diameters required in a water system for various levels of fire flow and peak hour conditions. The economic and water quality consequences of these design choices have been evaluated.

The methodology has been demonstrated by applying it to a realistic, though hypothetical, distribution system. System requirements for four different scenarios were evaluated:

1. Existing pipe and storage sizes.

2. Modified (downsized) pipes and storage that still will meet the full fire-flow requirements.

3. Modified pipe sizes and system storage to meet a reduced fire-flow requirement associated with having sprinkler systems in all structures.

4. Modified pipe sizes and system storage with no fire-flow requirements.

The adequacy of the various system design options was evaluated in terms of their capability to meet fire-flow requirements and to satisfy normal demands. Each option was examined in terms of its ability to meet maximum day conditions with a fire-flow requirement superimposed, and to meet peak hour conditions. The analysis includes both a steady-state assessment and an EPS that tested the ability of the system to deliver flow over the duration of a fire ranging from 2 hours to 4 hours. Subsequent to the identification of the system configurations under the four design options described above, additional modeling was performed to: 1) determine the cost of the resulting networks; and 2) determine the water quality consequences of the reduced pipe diameters and storage requirements.

In order to make the results more generally applicable, the same methodology was applied to a generalized network using four different design criteria, and three different fire-flow requirements, under three different population densities. A series of observations result from this analysis concerning the impact of these assumptions on pipe diameters, pipe costs, water age, and velocities.

Review of National and International Criteria and Providing Fire Flows

Water companies throughout North America and Europe, whether private or publicly owned, operate within a regulated environment and strive to provide the highest levels of service to their customers at levels of costs that reflect good value for the money invested in the water treatment and distributions systems. (Water UK and LGA, 1998) In terms of water supply for domestic purpose, the highest level of service usually means to provide a water service to customers at acceptable minimum flow and pressure, as well as supply water that meets applicable water quality regulations and is aesthetically pleasing. Water requirements for fire suppression or installed fire protection, on the other hand, are usually added to the requirements for domestic purposes and are commonly calculated according to discrete areas belonging to different fire-risk categories based on type of construction and building size. As discussed in greater detail below, fire-flow requirements vary greatly among countries. These variations may be attributed to the differences in construction materials and structural types that exist among countries. The following review is of countries that provided literature on water supply requirements.

United States

Throughout the United States, it is generally left up to the local authorities to determine the fire-flow requirements for their jurisdiction. The fire-flow requirements are sometimes specified in the model building codes or fire codes; however, if the political jurisdiction wishes to determine its own set of fire-flow requirements, there are three methods reported in the literature from the American Water Works Association. Each of these methods is identified below without full elaboration, for two of the methods are discussed in other referenced Chapters, and a third method has not been adopted by any standards or codes. Basic details are provided so that these methods can be compared and contrasted to other countries.

◊ The ISO NFF method: Refer to Chapter 6. The ISO method is used to determine NFFs for insurance rating purposes. This is a recommended practice for communities, but there is no legal requirement that political jurisdictions meet NFF criteria. However, municipal water supplies that do not meet these criteria may find that their commercial property owners are paying higher insurance rates; it may have less impact on residential property owners. The ISO method for calculating the NFF considers the construction (Ci), occupancy (Oi) exposure (Xi), and communications (Pi) factors of a building or fire division. Exposure and communication factors reflect the influence of exposed and communicating buildings on the NFF. (ISO, 2001) The ISO method is most likely to produce the most realistic requirements. (Farrell, 1996) NFF is calculated by the formula: $NFF_i = (C_i)(O_i)(X + P)_i$.

◊ Illinois Institute of Technology Research Institute (IITRI). The IITRI method is based on data collected from actual fires in several types of commercial occupancies in Chicago, and to determine the water application rate necessary to **control** a fire as a function of the fire area. Using a curve-fitting analysis, the IITRI method developed two equations solely based on flow area. The method has not been adopted by any code or standards-making organization.

◊ Iowa State University (ISU) method. The ISU method for calculating a required fire flow addresses both the quantity of water required to extinguish a fire, and the effects of various application rates and firefighting techniques. The equation used for this method is based on the consumption of fuel being dependent on the available oxygen supply in a 90-percent closed compartment or space, and vaporization of applied water into steam. (AWWA, 1998) A limitation of this equation is a result of the assumption that the entire compartment or space be entirely involved in fire. Therefore, the critical consideration involved in applying the ISU method is how a fire department fights structural fires. (Burns and Phelps, 1994) This method is further discussed in Chapter 6 and it should be noted that this method is fully explained in National Fire Protection Association (NFPA) 1142, *Standard on Water Supplies for Suburban and Rural Fire Fighting.*

In addition to these basic methods, building codes and standards in the United States also specify water-supply requirements for fire protection. The following building code information was identified from a review of the literature.

◊ **BOCA**: The BOCA National Fire Code references the NFPA codes and standards for the determination of required fire flows.

◊ **ICB0**: The Uniform Fire Code defines fire flow as the flow rate of a water supply, measured at 20 psi residual pressure that is available for firefighting. (ICBO, 1997) For one- and two-family dwellings, a fire area that does not exceed 3,600 square feet, the minimum fire-flow requirements are 1,000 gallons per minute (gpm). For other types of construction, the minimum required fire flow and flow duration for buildings are indicated in **Table 7-1**.

◊ **NFPA**: NFPA leaves it up to the authority having jurisdiction (AHJ) to determine the required fire flows for the local conditions in the form of local codes, ordinances, or resolutions. These local codes may, however, specify the minimum water supply requirements that must be available for firefighting purposes in areas where adequate and reliable water supply systems for firefighting purposes do not exist. The fire department having jurisdiction should perform an onsite survey of all buildings, including types of construction, occupancies, and exposures within the applicable jurisdiction to obtain the information needed to compute the minimum water supplies required. (NFPA, 1999b) NFPA 1142 provides calculations for determining the minimum firefighting water supply for structures based on their exposure hazard, total area of the structure, occupancy hazard classification number, and the construction classification number. Tables specify the minimum water supply to be 2,000 gallons and the maximum supply of 5,000,000 gallons that must be available in storage for firefighting purposes.

Table 7-1

Minimum Required Fire Flows and Flow Duration

Fire Area in Sq. Ft.*	Fire Flow in gpm	Duration in Hours
3,600 - 70,900	1,500 - 2,750	2
13,400 - 128,700	3,000 - 3,750	3
23,300 - 295,900 and greater	4,000 - 8,000	4

Source: ICBO 1997

*Fire areas for specific types of construction are provided in Appendix III-C; Fire Hydrant Locations and Distribution of the UFC (ICBO), 1997.

Canada

It is the responsibility of the provinces and territories throughout Canada to determine fire-flow requirements. The following is a summary regarding some of the codes and guidelines that are used by local authorities in Canada for determining the required fire flows for their province or territory.

- **National Building Code (NBC)**: The NBC does not dictate the required fire flow, but states that an adequate water supply for firefighting should be provided for every building. (Farrell, 1996)

- **Alberta Building Code (ABC)**: In 1973, Alberta became the first province to adopt the NBC by reference, making some amendments as noted in the ABC. (Farrell, 1996) The ABC provides a formula for calculating the supply of water to be available for firefighting purposes. The formula is based on the total building volume, a water supply coefficient, and a spatial coefficient. The water supply coefficient is selected from a table provided in the ABC, while the spatial coefficient is equal to either 1.5 or 1.0, based on the limiting distance between buildings. The water supply should be capable of being delivered at a rate of not less than 700 gpm for any building required to have a total quantity of water less than 20,000 gallons and a rate of not less than 950 gpm for a building requiring a quantity not less than 40,000 gallons. (Farrell, 1996)

- **Fire Underwriters Survey (FUS)**: Similar to the Insurance Services Office, Inc., in the United States, the FUS has developed a Guide to Recommended Practices and is responsible for performing the review and evaluation of the adequacy of fire protection for municipalities and rating them accordingly. The FUS requirement for fire flow is based on building characteristics including the type of construction, floor area, number of stories, nature of occupancy, automatic sprinkler protection, and exposure risk. The minimum recommended fire flow is 500 gpm, and the maximum recommended fire flow is 12,000 gpm. (Farrell, 1996)

- **Water Supply Standards**: The water supply standards for the City of Edmonton, Alberta, Canada, require a separate fire flow of 8 gpm to be imposed at a node adjacent to all high-value properties, e.g., schools and shopping centers. A separate fire flow of 1,600 gpm is also imposed at a node of the weakest link or farthest from the source of supply within the network for a single-family residential development. The residual pressure in all cases for any node within the network should not be less than 20 psi at ground level. (Farrell, 1996)

Europe

There are no fire protection standards that apply to all European countries; therefore, each country must develop or adopt its own fire-flow requirements.

Germany: The German standards for fire water requirements (DVGW, 1978) are based on very detailed fire-risk categories that assign the level of risk according to four classes of building use and three classes of fire spread risk (i.e., low, moderate, and high). German standards by building use categories are presented in **Table 7-2**.

Table 7-2

German Standards by Building Use Categories

Category	Description	Number of Stories	Floor Area Ratio*	Volume-Area Ratio#
1	Small building	≤ 2	≥ 0.4	C
2	Residential, light industrial, mixed residential, business mixed	≤ 3	0.3-0.6	C
3	Central business district, light industrial	3	0.7-1.2	C
4	Central business district, heavy industrial	1	1.0-2.4	_ 9

Source: adapted from DVGW 1978
* Floor Area Ratio = Total area of building/Footprint area
# Volume-Area Ratio = Total building volume/Footprint area

The combined risk categories result in a fire-flow requirements ranging from 200 gpm for the lowest risk area to 850 gpm for the highest risk area. The German standards require that all fire-risk flows should be maintained for at least 2 hours with the residual fire pressure (the minimum pressure required during fire-flow conditions) of 20 psi for all categories of risk.

United Kingdom: The flow rates that the UK fire services ideally require to fight fires are based on the national document on the provision of water for firefighting. (Water UK and LGA, 1998) Similar to German standards, fire-flow requirements are determined based on the categories of premises to be protected. **Table 7-3** shows UK fire-flow requirements ranging from 120 gpm for the lowest risk areas to 1,200 gpm for the highest risk areas.

The UK guidance document does not stipulate the period for which the required fire flow for a specific risk category should be maintained, or the residual fire pressure.

Table 7-3

Categories of Premises and Fire Flows as Required in the UK

Category	Description	Fire Flow in gpm
1	Housing	
	Not more than two floor	Minimum of 250
	More than two floors	300 to 550
2	Transportation (lorry/coach parks, multistory car parks, service stations	396
3	Industry:	
	Up to 1 hectare	300
	1 to 2 hectares	550
	2 to 3 hectares	800
	Over 3 hectares	1,200
4	Shopping, office, recreation, and tourism	300 to 1,200
5	Education, health, and community facilities:	
	Village hall	250
	Primary schools and single-story health centers	300
	Secondary schools, colleges, and large health and community facilities	550

Source: Adapted from Water UK and LGA, 1998.

Greece: The Technical Chamber of Greece (1996) stipulates that fire flows should be maintained for at least 30 minutes at a level that ranges from 200 gpm to 1,900 gpm, depending on the risk category. The required residual pressure, which is 60 psi for all categories, indicates that the main rationale behind the residual fire pressure in this case may not be to act as a safety factor to prevent negative pressures developing in the water supply network (like in other countries), but to supply pressure directly to fire hoses.

France: The French standard for fire water requirement (Circulaire 1951, 1957, and 1967) requires flows of 260 gpm supplied for 3 hours as a minimum. Similar to other European standards, fire-flow requirements are determined based on the risk associated with premises to be protected, and in case of increased risk (e.g., factories, apartment buildings, etc.) the requirement will increase accordingly. The level of risk and required fire flows are determined based on assessment by the firefighting services. As a consequence of this approach, more stringent requirements can be imposed. (Ruellan and Tiret, 1990) Such an example is given for the City of Paris, where the requirements range from 530 gpm for a house with not more than two floors, to 2,640 gpm for industrial zones of importance to the general public. Although there were attempts in 1977 and 1978 to re-evaluate these standards, the standards described were still in use in France in 2002.

Spain: Similar to the French standards, the Spanish standards for fire water requirements (Real Decreed, 1942/1993) require flows of 260 gpm to be supplied over 2 hours. The required minimum pressure is 13 psi.

Netherlands: In the Netherlands, water requirements historically are calculated based on now-outdated KIWA Mededeling 50 (KIWA 1977) according to defined risk areas: a) the high-risk fire flows (required provision of 1,600 gpm to be supplied over 6 hours), or b) the low-risk fire flows (400 gpm to be supplied for 2 hours), which are the highest amount the European countries defined in the subject report. The required fire pressure specified in this document is 30 psi for all categories of risk. However, 10 to 20 percent of the higher risk buildings still will require fire flows of 260 gpm. These higher risk standards are considered to be more of an exception than a rule. In the Netherlands, the fire department typically relies on the water utility for the first 15 to 30 minutes of a fire, after which time they use nonpotable water sources such as canals.

The Identification of Alternative Methods and Technologies for Structural Fire Suppression

Automatic Suppression Technologies

A host of structural fire suppression technologies that could have a positive impact on reducing water demand for fire structural suppression were reviewed by both committee and supporting staff. The following is a synopsis of the technologies presented in the document under review for this manual. However, there has been significant advancement in several of these technologies, along with better performance indicator since the AWWA Research Foundation/KIWA report was published in 2002. Therefore much of the detailed information is to be presented in the report's Chapter 9: Viable Approach to Providing Adequate and Reliable Water Supplies for Fire Suppression and Chapter 10: The Latest Concepts on Water Supply Systems.

Automatic sprinklers: Automatic fire sprinkler systems were the first and still are the foremost form of automated fire suppression system used throughout the world. With over 100 years of operating history and a 95-percent-plus success record, these systems represent one of the most readily available means for effective fire suppression for a wide range of different occupancies. The success of automatic sprinkler systems in confining, controlling, and extinguishing developing fires in structures has been phenomenal.

When used in a closed-head sprinkler system, each individual fire sprinkler is, in a sense, its own fire detector and suppression device. A fire sprinkler is constructed of a metal frame with a threaded pipe inlet and a water distribution deflector; on the interior of the threaded inlet there is a discharge orifice through which the water discharges; a cap covers the discharge orifice. This cap is held in place by one of several types of heat-responsive actuation mechanisms, (e.g., fusible link, chemical pellet, frangible bulb, or a special heat-responsive element that opens and closes the orifice based on the heat intensity. The distinction between the different types of actuation mechanisms is not significant relative to the sprinkler head's basic operation, only to the sensitivity of the operation. Each actuation mechanism has a predetermined (based on hazard to be protected, and as defined by code) temperature set point. Once the mechanism has been heated to this set point, the mechanism will fuse, melt, rupture, or otherwise open the orifice, boosting the water pressure in the piping system (in a wet-pipe system) allowing water to discharge in a defined pattern over the fire.

Fire sprinklers have orifices ranging from 0.25-inch diameter up to 1.0-inch diameter, depending on the type of hazard or occupancy classification to be protected. Temperature actuation set points range from a low of 135 °F (57 °C) up to 650 °F (343 °C). Set points also are selected based on occupancy. Sprinkler water flow rate is a product of the square root of the water pressure at the sprinkler orifice multiplied by a K factor representing the sprinkler discharge characteristics. K factor ranges are from 1.3 up to 25.2 (and even larger for large drop-type sprinkler heads). The standard sprinkler as defined by NFPA 13, *Standard for the Installation of Sprinkler Systems*, has a 1/2-inch (0.5) diameter orifice and a K factor of 5.6 and an actuation temperature of 155 °F (68 °C). At 15 psi a standard open sprinkler head delivers just under 25 gpm.

Sprinkler systems are designed and arranged with special control valves that include wet-pipe systems, dry-pipe systems, pre-action systems, and other special, less frequently used, systems. Wet-pipe sprinkler systems are by far the most common type of sprinkler system in service in commercial and residential properties today. Wet pipe, as the name implies, means that the piping distribution system to each sprinkler head is charged with water under pressure. It is important to note that in this type of closed-head system, each sprinkler fuses individually in response to heat input from a fire. Only those sprinklers directly exposed to the heat of the fire will actuate to spray downward onto the developing fire. A common misconception is that, when a sprinkler system actuates, all heads open simultaneously; this only happens with deluge systems, which are installed for very special hazardous situations. System actuation is alarmed through instrumentation at the system riser fed by a lead in water main that detects water flow into the system and is wired to a local remote fire alarm panel.

Dry-pipe sprinkler systems operate similar to wet-pipe systems except that the piping distribution system to feed the sprinkler heads is charged with air or an inert gas under a specified pressure. When an individual sprinkler head fuses, air or gas is discharged initially until the air pressure is depleted to a point where a differential type check valve located in the system riser will open and allow water to flow into the piping distribution system. Again, water flow is monitored by means of instrumentation to sound an alarm immediately. Dry-pipe systems typically are installed for the protection of areas where the sprinkler distribution piping is located in an unheated space such that, if water were present, it has the potential to freeze and damage the piping, resulting in undesirable non-fire-related discharge of water. The design water flow requirements for dry-pipe sprinkler system are basically the same as for wet-pipe systems except for the main control valve arrangement.

Pre-action sprinkler systems are used in areas where non-fire-related water discharge (e.g., due to damage to a sprinkler head or sprinkler piping resulting in an equipment failure) is highly undesirable, resulting in water damage, such as to a rare book collection. This system combines the design features of the dry-pipe sprinkler piping system and is charged with air or, more likely today, with an inert gas with a separate fire-detection system in the protected space. Special pre-action control valve is located at the system riser and is tied to the fire detectors in the protected space. The pre-action valve will not open to release water into the system until a fire alarm signal is received from the fire detectors. The signal will release the valve to admit water into the distribution system piping, but water discharge will **still not** occur until the heat of the fire has fused the actuation mechanism of an individual sprinkler head. Repeating, but this point is most important, water will only discharge from the sprinklers that have been heated to their actuation set pont temperature. When the system is in a standby (nonfire situation) condition, the pressure of the inert gas (i.e., air) in the piping is monitored by a low-pressure switch. If a low-air-pressure condition occurs due to pipe damage or sprinkler head damage, a trouble alarm will sound so that repairs can be made prior to any water discharge.

Deluge sprinklers and water spray systems: Deluge sprinklers and water spray systems operate in a similar manner to the pre-action systems described above. The difference in these types of systems is that the actuation mechanism in each sprinkler head has been removed so that all heads open on the system, and water will discharge simultaneously. In both types of systems, fire detection devices are placed throughout the protected space and will actuate a normally closed deluge-type valve at the system riser to flood the distribution piping and discharge water from all sprinklers or spray nozzles. Deluge sprinklers and water spray systems differ only in that deluge sprinklers are standard-type sprinkler heads and are used for general floor area protection the same as other types of sprinkler systems. However, water-spray nozzles are of a different design where the water pattern is directed at 30, 60, or 90 degrees from the nozzle orifice and are used for specialized applications such as the oil in transformers or to protect exposed structural steel. Both deluge sprinkler systems and water-spray systems typically are limited to use in high-challenge, rapidly developing fire hazard applications. Deluge systems may be used to protect the storage of piled combustibles such as bailed cotton, cardboard containers, or containers of flammable and combustible liquids. These types of systems are for special hazard protection only. They are inappropriate for use in other types of occupancies due to water damage potential.

Foam-based sprinklers: Wet-pipe, dry-pipe, pre-action, deluge, and water-spray-type fire protection systems, when used for the protection of highly challenging type fire hazards that include flammable liquid hydrocarbons, can be enhanced the addition through injection of one of several types of firefighting foam agents. The current basic foam classes include protein foams, fluoroprotein foams, and aqueous film forming foam (AFFF) for flammable liquid fires, and synthetic surfactant foams, better known as Class A foams, for ordinary combustibles and structural firefighting. These foams may be used with fixed system applications and also are commonly used with manual hose streams and master stream applications in high-hazard facilities.

The important concept here is that all of the firefighting foams are reported to extinguish fires faster and with less water in ordinary combustibles than plain water alone; this reduces the water demand for firefighting. Foam in general tends to form a foam blanket that helps exclude oxygen and reduce vapor emissions from the fuel surface, while allowing water droplets to settle out of the emulsion and provide cooling at the flame-fuel interface. While surfactant foaming agents are effective in controlling and extinguishing flammable liquid fires, and there is increased applications of the newer Class A foams for more rapid control and extinguishment of developing structural fires in residential occupancies, this technology does not have a sufficient history of carefully documented successes to consider the reduction of NFF at this time. Furthermore, the surfactant foams ultimately are carried to wastewater in the fire runoff. This may create an environmental concern regarding discharge of these soaps to wastewater streams in certain jurisdictions. (See further details on Class A Foam in Chapter 9.)

Residential and fast-response sprinklers: Within the past 2 decades, substantial decreases in the actuation time of sprinkler heads have been recognized. The design of sprinkler actuation mechanisms now incorporates a value known as the Response Time Index (RTI). RTI is a measure of thermal sensitivity, namely, the rate at which the sprinkler element can absorb heat from its surroundings before actuating. RTI incorporates the concept of thermal lag in that, while all actuation mechanisms are set to fuse at a given temperature, the mass of the actuating element introduces a lag time for heat flow into the element.

Significant research product development has led to two distinct types of state-of-the art sprinklers, both having faster response times: the residential sprinkler and the quick-response sprinkler. Both residential and quick-response sprinklers incorporate lower mass fusible links or frangible bulbs that decrease RTI. The difference in the two types is in their discharge characteristics. Residential sprinkler systems have a modified water discharge pattern, where more of the water discharged is contained in the upper portion of the horizontal discharge profile. This is necessary because residential sprinkler system design places a high reliance on the fire being suppressed with a single operating sprinkler head. The distribution pattern of the head must, therefore, protect draperies and higher elevation combustibles that are more typical to a residence than other types of occupancies. This discharge pattern also helps ensure that fire gases at the ceiling of an enclosure are cooled effectively, thereby reducing the likelihood of flashover and additional sprinklers actuating.

Water-mist technology: In the case of water-mist fire suppression systems, it might be observed that something 70 years old is now a new thing again. Water-mist technology was developed originally in Europe in the 1940s as application for marine engine and fuel space fire protection. In the 1950s, a significant interest and research was directed at water mist as one of the agents most viable for replacing the Halon 1301 fire extinguishing agent. Federal regulations stopped the use of Halon 1301 by 1980 as a significant threat to ozone depletion in the atmosphere. Beginning at the turn of the 21st century, significant interest and research has been directed again at water mist as one of the agents most viable for replacing Halon in certain applications.

Water-mist systems are defined by the different methods in which the mist is formed and the characteristics of the droplet size and distribution of the discharged mist from nozzles. Water-mist droplets are formed through specially designed nozzles at pressures typically higher than those of standard sprinklers. The pressure is generated by high-pressure pumps or accumulators, usually containing air or nitrogen, or even a combination of both. Water droplet size and discharge momentum are critical parameters in system design.

Additional Methods of Reducing or Augmenting Water Supplies for Fire Protection

The AWWA Research Foundation and KIWA study on the impact of fire flows on municipal water distributions systems also considered additional topics: 1) Surface Water Sources and Private Water Tanks, 2) Use of Nonpotable Water Delivery Systems for Fire Protection, and 3) Continuing Sprinkler Technology. This study was completed in 2001 and published in 2002. Since that time there has been significant progress in these areas along with water-mist technology that has been supported by the United States Fire Administration (USFA) in partnership program with the ISO, the National Institutes of Science and Technology (NIST), the NPFA Committee on Suburban and Rural Water Supply, Standard 1142, and the Marriott Corporation. This updated information is presented in Chapter 9 of the report on Viable Approach to Reducing Needed Fire Flow or Providing Independent Water Supplies For Fire Protection.

Study Summary Statement of Importance

This referenced study provides the following conclusions based on the information through the year 2001.

Water-efficient fire suppression technologies suggest the use of less water than conventional standards. In particular, the universal application of automatic sprinkler systems provide the most proven method of reducing loss of life and property due to fire, while at the same time providing faster response to the fire and requiring significantly less water than conventional firefighting techniques. It is recommended that the universal application of automatic fire sprinklers be adopted by local jurisdictions.

STUDY PREPARATION BY:

1. Jerry Snyder, Arun K. Deb, Frank M. Grabluta, and Sandra B. McCammon; Roy F. Weston, Inc., West Chester, Pennsylvania, 19380

2. Walter M. Grayman, Ph.D., Consulting Engineer, Cincinnati, Ohio

3. Robert M. Clark, Ph.D., The United States Environmental Protection Agency, Cincinnati, Ohio

4. Daniel, A Okun, Ph.D., University of North Carolina, Chapel Hill, North Carolina

5. Scott M. Tyler, Duke Engineering Services

6. Dragan Savic, University of Exeter, Exeter, United Kingdom

PROJECT ADVISORY COMMITTEE

Note: Three committee sessions were held in the Philadelphia, Pennsylvania, area and numerous phone contacts were made over a 2-year period. Some of the following individuals prepared separate documents that were used in writing the final study report.

- Mr. Isaac Pai, Long Beach Water Department, Long Beach, California
- Mr. Jonathan Clement, Black & Veatch, Boston, Massachusetts
- Mr. William Kirkpatrick, East Bay Municipal Utility District, Oakland, California
- Mr. Jan Vreeburg, KIWA, the Netherlands
- Mr. Jeffrey Swertfeger, Cincinnati Water Works (CWW), Cincinnati, Ohio
- Mr. P. K. Tudor, Delaware County Water Company (DEL-CO), Ohio
- Mr. Lewis Rossman, U.S. Environmental Protection Agency, (EPA), for modifications made to the EPANET software
- Mr. Scott Mitchell, American Fire Sprinkler Association (AFSA)
- Dr. Harry E. Hickey, Professor of Engineering, University of Maryland (Retired) representing the Insurance Services Office, Inc.
- Dr. Tom Walski, Haestad Methods, Waterbury, Connecticut

Chapter 8: Monitoring Water Supply Consumption and Security Analysis

Overview

This chapter is divided into two parts that are inherently dependent on each other. Part I is concerned with the monitoring of water supplies needed to meet the demand requirements of municipal water systems from both adequacy and a reliability capability. Part II is devoted to considerations for securing municipal water systems so that any interruption is minimized or eliminated. These conditions range from electrical power failures to pumping equipment to destructive interruptions from a terrorist attack. Issues and programs to bring about the objectives presented in Part II are still under active program development by the Department of Homeland Security (DHS), other Federal agencies, State governments, and local governments.

Part I

Both Volume I and Volume II of this project series on municipal water supplies has identified three historical or predicted water demand rates for consumer consumption and the relationship that these flow rates have on meeting fire protection requirements. Since these rates affect water supply consumption, they are reviewed as follows: (Reference #1, pg. 159)

- **Average daily consumption or demand**: The average of the total amount of water used each day during a 1-year period.

- **Maximum daily demand**: The maximum total amount of water used during any 24-hour period. The Insurance Services Office, Inc. (ISO) bases its calculations on the highest demand during the previous 3 years. For the Public Protection Classification process, water as the result of water main breaks or other water system failures, is not considered in evaluating the maximum daily consumption (MDC) rate where records are carefully kept. However, such water losses can have an adverse effect on meeting both consumer consumption needs and water for structural firefighting.

- **Maximum hourly demand**: The maximum amount of water used in any single hour of any day in a 3-year period. It often is expressed in gallons per day by multiplying the actual peak hour use by 24.

The monitoring of water supply demand today requires far more than estimating practices, especially as it relates to water consumption every hour and possibly even more frequently where total demand rates exceed over 1,000 gallons per minute (gpm) delivered to the water distribution system by the water treatment plant. Ideally, water supply and demand requirements on any given water delivery system require "real time" or constant monitoring of the distribution system from the water system operators control room. The number of operators required in the control room and the amount and type of monitoring system(s) is dependent on the size of the water delivery system in terms of flow rates and auxiliary storage capacity in ground-level and elevated storage tanks.

Water Supply Graphing of Water Flow in Distribution Systems

Circular pen graphs are widely used to record water flow for small and medium-size water system. The graph paper can provide read-outs in gpm and/or million gallons per day (MGD) rates. Graph paper may be calibrated for 1-hour, 6-hour, 12-hour, or 24-hour intervals. More modern and large water systems use digital meters linked to printers that provide strip recording by the minutes, and graph the data on a computer interface. By whatever means used, these daily records are to be maintained on file for a 3-year period for identifying problems and identifying any changes in demand, by day of the week, week of the month, the month of the year, or on a yearly basis.

The senior water supply operator at the water treatment and distribution plant has responsibility to monitor water consumption closely and adjust water plant equipment to meet consumer demand. This individual also has the responsibility to monitor structural fire alarms and, if there is a detected increase in water demand, to boost flow and pressures to meet this demand as needed. The bottom line is that **no** fire hydrant on a municipal water supply system should drop below 20 psi residual pressure during a fire, if at all possible. If pressures drop below the 20 psi level, the water superintendent needs to notify the ranking fire officer at the fire site of the water supply condition.

The use of water supply graphing on municipal water supplies can be enhanced not only by providing graphing at the water supply plant but by having real-time graphs on both ground-level and elevated storage tanks, along with any auxiliary pumping stations. These graphs are essential to keeping the water systems balanced in relation to supply and demand for both water delivery rates and water pressures. Remotely controlled valves from the water plant can be used to maintain constant flows at required pressures throughout the water distribution system.

Supplemental field graphing also can be accomplished by monitoring in-line flow meters equipped with transponders to transmit flow data on primary and secondary feeders at specific locations on the water distribution system. This is extremely useful in the following ways:

◆ It immediately identifies any water main break including break-off hydrant stems. If there is destruction of a water main, it can be detected almost immediately.

◆ It can be used to provide the fire department with real-time fire-flow information in the event of working fires.

◆ It can be used to redirect flows during periods of the day to meet dwelling water demand, commercial water demand, industrial water demand, and special-use water demand.

◆ It is also very useful for maintaining the calibration of computer models linked to water systems. (See Chapter 9.)

Water Supply Graphing of Water Source Supplies

It is no less important to graph the quantity of untreated water that can be made available at any given time for domestic consumption, for dual water systems, or alternative water supplies. The first consideration given below is water supplies that will be treated for domestic consumption. As previously noted, untreated water supplies may come from one or more of the following sources to supply a domestic water supply system:

◆ wells;
◆ impounding reservoirs, both low-level and high-level;
◆ rivers and streams;
◆ lakes, both small and large;
◆ oceans where desalination processing is available (See Volume I, Chapter 10.); and
◆ cross-connections to other municipal water systems.

Knowing the amount of water in gallons or millions of gallons is critical to the constant evaluation of existing water supplies in relation to the demand for water supplies. There is a wide range of gauging and sensing equipment, devices, and recording equipment on the market for constantly evaluating both quality and pressure functions associated with each of the water supply sources identified above. Some of the key information that should be known to the water system operator at all times for each of the water sources identified are as follows:

- **Well sites**: Total depth of water in the well without pumping. The available drawn-down level at fixed or variable pump speeds that will provide a specified rate of flow and the duration of that flow.

- **Impounding reservoirs**: The available acre-feet of water, which translates into millions of gallons of water that can be used by a computer program to estimate the duration of flow at specified rates. This can be used to restrict the use of water to extend the duration period until the water supply is replenished.

- **Rivers and streams**: There are two issues with flowing water from rivers and streams: the level of water over the intake pipes at all times, and the rate of flow past the intake point at all times. In the case of streams that are stocked with fish, it is possible to contact the local office of the department of environmental conservation and determine in real time what the water flow rate is in feet per second or gpm at gauging equipment that may be located close to the intake source point without installing expensive equipment at the waterworks plant.

- **Lakes–both large and small**: Again there is the need to know the level in feet of water over the intake ports to the pipe that is feeding the waterworks plant.

- **Ocean water**: The criteria are covered as referenced above.

- **Cross-connection to adjacent water systems**: Both pressure sensors and flow meters need to be installed at the flow point from the adjacent system that will provide readouts for the plant operator receiving water from this arrangement. Flow test curves should be constructed every 6 months at the connect point from nearby fire hydrants to assure that the expected flow and pressure are maintained. Any increases or decreases affect the decisionmaking of the plant operator for the system receiving the water. This last statement is intended to imply that water may be received from either of the two systems that become a party to the cross-connection arrangement.

Monitoring the General Quality of Water Entering a Treatment Plant

In the current world situation it is essential that water not be contaminated before it enters a water treatment plant. Therefore a continuous sampler needs to be installed on all intake lines that will sound an alarm and immediately shut down intake valves or pump intakes when suspected foreign products or substances are detected in the intake water supply. Such questionable products and dosages are currently being identified by the Environmental Protection Agency (EPA) and DHS.

PART II SECURITY ANALYSIS OF MUNICIPAL WATER SYSTEMS (Reference #2, pg. 159)

Hazard Assessment

According to the President's Commission on Critical Infrastructure Protection, three attributes are crucial to water supply end users:

- there must be adequate quantities of water on demand;
- it must be delivered at sufficient pressure; and
- it must be **safe** to use.

Actions that affect any of these three factors can be debilitating for the infrastructure. The first two attributes are directly influenced by physical damage. The third attribute, water quality, is susceptible to physical events as well as the introduction of microorganisms, toxins, chemicals, and radioactive materials.

Terrorism can strike all the system components individually or in various combinations. Damages associated with most of the hazards identified in the American Water Works Association (AWWA) Manual 19 can result from terrorist activity, which can range from major events that cause severe damage and disruption of system operations, to minor incidents that may not affect normal activities. (Reference #3, pg. 159)

[Sidebar: The security analysis of public-sector water supplies has to be an integral part of the constant monitoring requirements for water supplies.]

A hazard is defined as any individual, group, or event that could destroy the facility, halt or suspend operations, otherwise threaten the public health, harm employees, publicly embarrass a utility, require a utility to expend a great deal of time or money, or cause general panic. In preparing the hazard (or threat) assessment, security assessment specialists generally look at who or what constitutes a threat and how the identified individual group or event could attack a particular system. It is important to consider all potential threats, because assuming the worst may result in overlooking more likely attempts.

Who Poses a Threat?

The terrorists' goal is to achieve notoriety for his or her cause. This can be through massive loss of life, as was the case in the plane hijackings of September 11, 2001. Beyond the obvious human tragedy, the net result of these actions includes significant economic impact on industry, the Federal government, and State and local governments as they scramble to provide additional law enforcement, National Guard, and if necessary, military presence. The key to terrorism's success lies not in the act itself but in the ensuing lack of confidence or feeling of insecurity for the citizens at large.

The actual goal of the terrorist usually is not associated directly with the target selected. The usual desire is to install a general sense of fear in citizens to disrupt their normal activities and way of life. Consumer uncertainty can significantly affect a national economy. Public perception of the safety of infrastructure, as well as the infrastructure itself, are both important assets to be protected. The following information describes who may pose a threat to public sector potable water systems.

◈ **Vandals**: A vandal often has a goal in mind, but not necessarily a target; thus the crime may be of opportunity. Some typical examples of such a crime are graffiti or broken windows. However, a vandalism problem can become much more serious if the paint used to write the graffiti on a reservoir dam is then dumped into the reservoir, posing a potential threat to the health and safety of the water system consumers.

◈ **Individual person**: An individual is someone working independently. Although the motivations of the individuals can vary widely, the target is clearly defined as the final water user or the facility itself. Individuals who purposely threaten infrastructure are often mentally ill and may target victims based on ethnicity, beliefs, or other characteristics.

◈ **Insiders**: With their detailed knowledge of the facility and water system, past and present employees and contractors pose some of the most serious threats to the disruption of water supply systems. Motivations of this group may include revenge or the venting of anger manifested from a real or imagined problem.

◈ **Domestic extremist groups**: Cults and extremist groups with a political agenda pose a threat to water systems. In the past, food has been contaminated with Salmonella with the objective of affecting voter turnout. The Aum Shinrikyo cult of Japan spent $30 million on a poisonous gas plant for use in terrorist activities. Better known for Sarin gas attacks, the cult also developed anthrax and attempted to use it as a weapon against civilian populations in downtown Tokyo in 2000.

◈ **State-supported terrorist organizations**: These organizations usually have a large number of followers and the greatest financial and technological resources. The use of some chemical and biological weapons of mass destruction are limited to these groups because of the significant resources required for their development. Known nations with the capabilities to produce weapons of mass destruction are North Korea, China, India, Pakistan, Iran, Iraq, Syria, Libya, and Russia as reported by the Department of Defense in 2001. (Reference #2, pg. 159)

Types of Threats

The follow information describes two types of threats against a water utility: physical and contaminant threats.

Physical Threats

Many observers believe that a physical event that destroys or disrupts a water system's components is a much more likely scenario than a contamination event. For instance, explosive materials are readily available and require a lower level of education compared to the development and deployment of contaminants. Potential types of physical attacks are

⬥ Aerial attack includes physical attacks on the treatment facility or the use of airplanes to drop contaminants into open reservoirs. (Note: The latter example combines a physical attack with a contaminant attack.)

⬥ Cyberterrorism attacks on the data acquisition system (supervisory control and data acquisition [SCADA]) are another threat. A terrorist could disguise the data, neutralize the chlorine, or add no disinfectant, thus compromising disinfection and allowing the addition of microbes usually not considered a threat, such as salmonella (when chlorine residual is present). Alternatively, attacks on the central control systems could create a large number of simultaneous main breaks by opening and closing major control valves too rapidly. Because many SCADA and control networks are not connected to the Internet, this threat is most likely to come from a disgruntled employee with access to the system.

⬥ Explosives could be used at any number of locations to compromise the pumping, storage, or transmission of water. Explosives, which can be developed or obtained, pose less risk to the attacker compared to biological and chemical weapons. Explosives also require a comparatively lower level of education. A bomb explosion within the distribution system will require immediate response and redirection of water to prevent contamination and draining of the water system.

⬥ Fire may be one of the easiest methods of sabotage because needed materials are readily available. It is effective because destruction of the computer control system, pumps, or motors or compromising the structure pose a significant barrier to the operation and timely restart of a water treatment plant. Furthermore, once the city water supply is reduced, the ability to fight other fires is compromised, seriously affecting the safety of other critical infrastructure elements.

⬥ Personal attacks on the plant staff could lead to multiple personnel injuries that would leave a plant without a skilled operational workforce. A hostile takeover also could allow for a cyberterrorism attack.

Contaminants

The following subject items outline biochemical toxins; microbial agents; industrial chemicals; nerve, blood, and blister agents; and radioactive materials that potentially could contaminate potable water systems. The chemistry of each toxin, chemical, and microbial agent is specific. Some are neutralized by chlorine, others are effectively removed through the drinking water treatment process, and all have different thresholds for the appearance of symptoms, infection, and lethality.

⬥ Biochemical toxins require a very small volume compared to other chemicals. Still, many are difficult to develop in quantities large enough to pose a lethal threat to municipal water systems; however, smaller, nonlethal doses may be used to induce sickness or terrorize the population. **Table 8-1** includes a list of potential toxins and their sources.

Table 8-1

Biological Toxins

Agent	Source
Abrin	Plant (Rosemary)
Aconitine	Plant (Monkshood)
alpha-Conotoxin	Cone Snail
alpha-Tityustoxin	Scorpion
Anatoxin-A(s)	Blue-Green Algae
Batrachotoxin	Arrow-Poison Frog
Botulinum toxin	Bacterium
C. *pertringens* toxins	Bacterium
Ciguatoxin	Marine Dinoflagellate
Diphtheria toxin	Bacterium
Maitotoxin	Marine Dinoflagellate
Microcystin	Blue-Green Algae
Palytoxin	Marine Soft Coral
Ricin	Plant (Castor Bean)
Saxitoxin	Marine Dinoflagellate
SEB (Rhesus/Aerosol)	Bacterium
Shiga toxin	Bacterium
T-2 toxin	Fungal Myotoxin
Taipoxin	Elapid Snake
Tetanus toxin	Bacterium
Tetrodotoxin	Puffer Fish
Textilotoxin	Elapid Snake

Source: U.S. Army Medical Research Institute of Infectious Disease, 2001.

🌢 Microbial agents include bacteria, virus, protozoa, and other microbes. Experts and the U.S. government believe biological weapons are within the reach of terrorists. However, the education level, monetary resources, and risk required to produce these agents are higher than those required for physical methods of terrorism. The high microbial concentration place the developer of such weapons at high risk. **Table 8-2** provides a partial list of potential bacterial and viral agents.

Table 8-2

Potable Water Pathogens

Scientific Name	Common Name	Availability in Environment
Bacillus anthracis	Anthrax	Infected cattle, goats, swine, sheep, horses, mules, dogs, cats, wild, animals, and birds
Brucella melitensis & Brucella suis	Brucellosis	Infected cattle, goats, swine, sheep, horses, mules, dogs, cats, fowl, deer, and rabbits
Vibrio cholerae	Cholera	Human excrement and shellfish
Clostridium perfringens	Clostridium Perfringens	Soils, water body sediment, intestinal tracts of fish and mammals, crabs, and other shellfish
Cryptosporidium parvum	Cryptosporidiosis (Crypto)	Calves
Encephalomyelitis Virus	Encephalomyelitis (VEE)	Rodents and horses
Picornaviridae & Reoviridae	Enteric Viruses	Humans
Burkholderia mallei	Glanders	Horses
Yersinia pestis	Plague	Prairie dogs, chipmunks, black rats, deer mouse, species of ground squirrels, and coyotes
Chlamydia psittaci	Psittacosis	Birds
Coxiella burnetii	Q Fever	Cattle, sheep, and goats
Salmonella typhimurium	Salmonella	Fowl, swine, sheep, cattle, horses, dogs, cats, rodents, reptiles, birds, and turtles
Shigella dysenteriae	Shigellosis	Sewage
Variola Major & Variola Minor	Smallpox	Centers for Disease Control and Prevention; Russian Biological Lab
Francisella tularensis	Tularemia	Wild rabbits and most other wild and Domestic animals
Ebola & Hantaviral among others	Viral Hemorrhagic Fever (VHF)	Ticks and rodents

Source: Prescott, L.M., J.P. Harley, and D.A. Klein, 1999.

🌢 Industrial chemicals are yet another threat due to large, readily available supplies. Several factors are important in the analysis of a chemical threat: volume of water to be contaminated, solubility of contaminant, lethal dose, and volume of water that must be ingested to constitute a lethal dose.

Fortunately, the vast majority of industrial chemicals make poor candidates as a lethal, undetectable agent. Often the lethal dose required to contaminate a water supply requires a very large quantity or even is insoluble at the required concentration. Furthermore, many toxic chemicals have disagreeable colors, tastes, and odors that would alert the consumer to their presence.

Although poisoning the water supply through the use of industrial chemicals is difficult, it is not impossible to make the water unfit for consumption.

🝆 War agents such as nerve, blood, choking, and blister agents include sulfur, mustard, and Sarin gases among many others. These have been developed by many countries for use, generally in incapacitating or discomfort agents. They are not considered as likely toxic water contamination threats due to the high concentrations required.

Nerve agents are the most deadly within this category, as they are 100 to 1,000 times more lethal than pesticides made with organophosphorous chemicals. (Smithson, 2000)

🝆 Radioactive material is another method of contamination. The primary radiological threat is the use of conventional explosives to spread radioactive contamination over a limited area of strategic terrain. This could include highly radioactive materials, such as spent fuel traveling to the Yucca Mountains for containment or low-level radioactive materials, including uranium-238, iradium-192, cesium-137, strontium-90, or cobalt-60. Using radioactive materials to contaminate drinking water presents a challenge, requiring large quantities of materials, many of which are insoluble in water, heavy, and would settle out before reaching the target or would be trapped in filters. Furthermore, radioactive material poses a safety concern for the attacker.

Vulnerability Assessment

The following material adds consideration for terrorist activities to the four basic steps in a vulnerability assessment.

1. Identify and describe the separate components of the total water supply system.

2. Estimate the potential effects of probable disaster hazards on each component of the system.

3. Establish performance goals and acceptable levels of service for the system.

4. If the system fails to operate at desired levels under potential disaster conditions, identify key or critical system components responsible for the condition.

Step One: Identify Major System Components

Key elements of the total system should be listed and described as components under the following general headings: 1) administration and operations, 2) source water, 3) transmission system, 4) treatment facilities, 5) storage, 6) distribution system, 7) electric power, 8) transportation, and 9) communications.

Describe system components with as much detail as possible. Typical items included in a general description are pressure zones, location of pressure relief valves, pipe sizes, pipe materials and ages, typical distance between fire hydrants, and **all** system control valve locations.

In addition to these physical components, an additional item to be considered is the public confidence and reputation of the utility. An important asset of water treatment and distribution systems is the public perception and confidence in the end product. This also must be considered vulnerable to terrorist attack.

Step Two: Determine the Effects of Probable Disaster Hazards on System Components

The effects of a terrorist event on a water utility can result in a wide range of consequences. For example, an explosive device detonated at a noncritical location may not cause any appreciable damage to the facilities and therefore not endanger the capability of the facility to process water. Conversely, chemical contamination of the system can result in long-term disruption of service until it can be cleaned and returned to service. The range of consequences that could be attributed to terrorist activities include

- disruption of water treatment, storage, and delivery, and delivery components;
- introduction of biohazards and toxins into the water system;
- injury to facility personnel;
- injury to the general public;
- damage to utility property or equipment;
- damage to private property; and
- hazardous waste disposal problems: i.e., what happens to contaminated water if it is flushed from the system?

In addition, terrorist activity may focus on more than one part of the system, using damage in one area to divert the response team or to magnify the consequences of damage to another element of the system.

Step 3: Establish Performance Goals and Acceptance Levels of Service for the System

A water system is considered a lifeline because water is essential to the safety and health of the population it serves. A utility should develop specific goals and acceptable levels of service under disaster and recovery conditions. The acceptable goals for system service should consider the effects of terrorist activity. Taken individually, the effects are identical to those caused by various natural and human-induced hazards. Specific goals to consider are life safety, fire suppression, public health needs, and commercial and business uses.

Step 4: Identify Critical Components

Identifying the critical components of the system or its subcomponents is the final step in the vulnerability analysis. Critical components are those vulnerable to failure or partial failure because of an intentional act or natural disaster. Failure of a critical component will reduce the system's ability to meet minimum health and safety performance goals. To identify those components that would fail in an intentional attack, run a desktop exercise of an attack scenario and then focus on those components whose failure would render the entire system inoperative; these are the most vulnerable components.

Consideration of critical components should include the public perception of the value of a safe water system. In most cases, confidence in the safety of the water system and financial support are closely linked. The public most likely will choose to provide support for security measures if it is perceived that a well-conceived security master plan is in place.

Critical Review of Existing Security System

The purpose of security systems is to limit vulnerability. It is important to assess existing security measures and their integration with previous steps to determine the effectiveness of these systems. Do the existing security measures protect major and critical system components and do they minimize risk? The security review should include facility inspection, a review of documents and operations (policies, plans, and Standard Operating Procedures (SOPs)), and interviews with employees.

MITIGATION

Mitigation actions to reduce system vulnerability to terrorism are very similar to methods discussed at length in other publications about mitigating vandalism or natural hazards. However, mitigation actions also include elements that prevent unwarranted access to a system's components, which in some cases may be contrary to the ease of access (e.g., reaching a water intake by boat) that is necessary for mitigating an unintentional disaster. The extent to which these measures are applied will depend directly on the acceptable amount of risk and the likelihood of the hazard materializing, which are determined by the hazard analysis and vulnerability assessment.

Mitigation at the Source

Standard mitigation measures relating to source water and transmission range from providing alternative sources and protecting wellheads to retrofitting dams or aqueducts. Mitigation actions for watershed damage or widespread contamination include providing automated monitoring equipment, using alternative sources or intakes, and modifying source water treatment at the plant. Controlling access, identifying alternative sources, and providing flexible treatment facilities also can mitigate the effects of deliberate contamination of reservoirs. Access to reservoirs and to other outlying system components, such as pump houses and tanks, can be controlled by installing fences, gates, and signs; closing unnecessary roads; and increasing security patrols.

Preventing Access to Facilities

To prevent access to facilities, install adequate locks, window security, and lighting. Install intrusion-prevention devices, such as electronic keys, identification-card checkers, and 10-key code units, to control access to the strategic facilities. If a facility has the personnel to perform constant monitoring or can contract with a security firm, one also can install closed-circuit television monitoring systems or alarm systems with ultrasonic, heat, or beam sensors and magnetic switches to detect intruders. Be sure to change passwords when employees are dismissed or a contractor's job is done, both on electronic keys and on computer systems.

Water supply superintendents should not forget about controlling access to chlorine and other chemical systems, and design and construct these systems with automatic control systems that can indicate the extent and location of a leak, actuate chlorine scrubbers, close valves, shut down equipment, and isolate affected areas.

Distribution System Issues

Because of their large numbers and widespread dispersal, controlling access to distribution system components can be difficult. One key to accessing or locating any attacks on the distribution system will be accurate and up-to-date system plans, which should be maintained in numerous locations.

Water department personnel need to work with the local fire departments to be sure there is adequate alternative pumping capacity and water supply in the event of loss of flow to fire hydrants in an emergency situation. Consider what action should be taken to provide these redundancies and where the water would come from if a storage tank were drained or a main break interrupted the flow.

Mitigation activities should take into account what happens to contaminated water if it needs to be flushed from the system. Does exposing it to the air cause even more potential for dispersal of the contaminant? Is the water normally released to a storm drain or waterway?

Staff Role in Preventing Terrorism

In addition to the mitigation techniques identified, facility personnel also can play a very important role in preventing terrorism and acts of sabotage to facilities. In the wake of recent terrorism actions, a new sense of patriotism has emerged. Facility personnel who previously may have been hesitant to take an active role in security activities generally now are eager to be involved in facility security. Most Americans recognize that they can play a role in protecting citizens at home. It is most important to include the front-line employees when discussing what could happen and what action to take; these people are the eyes and ears of the system and may be the first to spot trouble.

Water utilities should be much more careful in disseminating information regarding facility operations, plant and system layouts, and emergency response and crisis management plans. These could prove useful to a terrorist wishing to identify system vulnerabilities.

Two factors that will reduce water system susceptibility are the reduction of information regarding facility operations and preventing access to the target through countermeasures. First is the reduction or control of information available to the public. Information concerning the plant and distribution system layouts and emergency response and crisis management plans could help terrorists identify system vulnerabilities. These sources of information should be removed from public libraries, the Internet, and other available locations. Second, those who claim to need legitimate access to such documents should be scrutinized.

Countermeasures

The following items are considered effective countermeasures against terrorism placed in the context of municipal water supply systems:

- Access control can include a variety of systems designed to control the movement of persons or vehicles. Access control may include guards, locks and keys, or access cards.

- Physical barriers may have been established to prevent an intrusion or attack, but usually function to hinder or slow the attack. Such measures may include hardened construction, vandal-resistant glazing, and doors. **Fencing** is not a good barrier, as it is only considered to provide a 6- to 10-second delay in access to the perimeter of a facility. Physical barriers in regard to contaminants may include back-flow prevention, filters for airborne particulate on plant and reservoir vents, or specially designed reservoir vents to prevent the pumping of a contaminant into the distribution system.

- Detection through some type of monitoring system provides notification that an undesired intrusion (physical or contaminant) has occurred. Examples include name badges to identify authorized personnel, closed-circuit camera, motion detectors, security guard scrutiny, or sensors in the distribution system. Detection should be as far away from the asset as possible to allow time for the response. Closed-circuit television systems make excellent assessment devices. Assessment refers to the verification of the detection system. For instance, personnel can use monitors to help assess the situation after an alarm goes off.

Once the utility decides which countermeasures to implement, any physical improvements should be combined with updated policies and procedures to ensure full optimization of the system.

RESPONSE PLANNING FOR PUBLIC DRINKING WATER SYSTEMS TO MAINTAIN WATER QUALITY

All elements of a water treatment plant are susceptible to human disruption, including the raw water source, the treatment facility, the operations and control facility and systems, and the support facilities and system such as chemical feed, power supply, and communications equipment. A preparedness plan and a response plan can be developed based on the results of the vulnerability and the risk of any given water utility. A response plan must define the measures that will be implemented to minimize the likelihood of an event or to mitigate its impacts.

According to the Studies in Urban Security Group (SUSG) at the College of Architecture and Urban Planning in Ann Arbor, Michigan, the development of an emergency response plan for the contamination of a public water system may be subject to Federal, State, or local regulations or guidelines (Rycus, Snyder, and Meier). Presidential Decision Directive 63 requires that Federal agencies develop and implement plans to protect the Nation's critical infrastructure. The Safe Drinking Water Act of 1986 and the Emergency Planning and Community Right-to-Know Act (EPCRA) of 1986 require that each State appoint a State Emergency Response Commission (SERC), whose responsibilities include designating emergency planning districts within the State.

Following the designation of emergency planning districts, Local Emergency Planning Committees (LEPCs) should be formed. These committees should consist of representatives of public agencies, such as water and wastewater utilities, fire department, health officials, law enforcement, and government officials. The LEPC then reviews the responsibilities of the water service during storms, floods, earthquakes, fires, explosions, nuclear reactor spills, aircraft crashes, hazardous materials incidents, power failures, and civil disturbance when formulating the response plan for human disturbance.

AWWA Manual M-19 provides utilities with information helpful in formulating a response plan. In addition, the SUSG study states that a response plan should contain (Reference #3, pg. 159)

- a legal and administration basis;
- a classification of emergency conditions; and
- provisions for command and control, communications, emergency supplies, distribution, threat management, and plan review and revision.

The emergency plan needs to contain the following updated listing of definitions and development phases which were revised after 9/11.

Define the Emergency Status

In developing a response plan, the term emergency must be defined, preferably with variations in degrees of alert and emergency status. The SUSG study defines four levels of severity as follows:

- **Normal operations/Minor emergencies**: Required responses do not extend beyond the water department, so an emergency response is not warranted.

- **Alert condition**: In situations where a major emergency may be forthcoming, the system director may declare an alert condition. An alert condition triggers the assembly of key decisionmakers and operational personnel to assess and monitor the situation.

- **Emergency condition**: In a situation where disruption or contamination is imminent or has occurred and where the full resources of the system, augmented by external resources (e.g., fire, police, public health), are required for appropriate response, the system director should declare an emergency condition. Under this condition, a full response plan would be implemented.

- **State of Emergency**: At this most serious level, especially involving the community at large, declaration of a State of Emergency would be appropriate. Typically, only the Governor of the applicable State can declare this condition, and it implies that the broadest resources available will be applied to the problem.

Developing an Emergency Response Plan

The development details are divided into nine general steps that should be considered in the development of a response plan for a public potable drinking water system. When developing a response plan for a specific water utility, it is important to evaluate each of the following steps, investigate the relevance of each step to that specific water utility, expand on the information and provide a documentation of the plan and all useful information.

Step 1: Gather the Command Group at a Designated Location

The SUSG study recommends that the Command Group include persons with specific expertise, such as the chief operator of the water utility, a staff member capable of providing technical support, staff member for administrative support, the head of engineering, the director of laboratories, and the director of security. The group should be organized according to an established chain of command.

The Command Group members should meet at a pre-arranged location, or Emergency Operations Center(EOC), that contains working communications equipment, including a computer with information on distribution system plans, source water plans, and secondary or tertiary utility plans, computer Internet capability, and one or more telephones and an emergency radio system. (Rycus, Snyder, and Meier) A backup location such as the local EOC also should be established in the event that the primary location cannot be used.

Step 2: Conduct a Preliminary Assessment to Determine the Nature, Extent, and Severity of the Disturbance.

A plan for assessing a disturbance should be prepared in advance. The assessment should address control around the perimeter of the utility's assets and an internal review of water treatment plant data encompassing the laboratory, water treatment plant performance, disinfection contact times, chemical feed, and other relevant data. The personnel who participate in the assessment should have an intimate understanding of plant operations and the significance of all data elements.

Water utilities are not required by the EPA to monitor their facilities for biological, chemical, or radiological contaminants or cyber interference of raw water sources or finished water supplies. However, as previously noted, such practices are recommended. Also recommended is the continual monitoring, routine testing, and careful observation of water supplies needed to prevent contamination of supplies or interference with operations and to facilitate prompt mitigation of any disturbance. The local department of health or the laboratory that performs the water analysis should immediately report any contamination or suspicious test results to a designated member of the Command Group, most likely the water utility supervisor. Any information from the laboratory will enhance the assessment.

One way to determine the threat posed if a chemical contamination occurs within the utility is a Computer-Aided Management of Emergency Operations (CAMEO) database. Many SERCs have access to a CAMEO database, which includes information on chemicals, transportation of the chemicals, and other information to better prepare for response to chemical emergencies. The EPA's Chemical Emergency Preparedness and Prevention Office and the National Oceanic and Atmospheric Administration (NOAA) Office of Response and Restoration developed CAMEO. The EPA also published a document about SERCs and the involvement of the CAMEO program titled *Secrets of Successful SERCs*. (EPA, 1993)

Based on the results of the assessment, the Command Group should decide whether to declare a water utility alert or a water utility emergency and, if necessary, implement the appropriate emergency response.

Step 3: Assemble Specialized Groups

Specialized groups should be assigned before the emergency occurs. This will save time and expedite the group's response actions. The assignment of the specialized groups may include

- situation assessment;
- laboratory analysis;
- law enforcement and security;
- public information/media communications;
- emergency water supplies;
- emergency evacuation;
- human health reporting/assistance; and
- repair/recovery.

A secured and current membership list for each group should include each member's position, name, and phone numbers. This list should also be included within the response plan.

Step 4: Alert Other Officials

If an emergency is declared, the Command Group should request further assistance from the appropriate law enforcement, fire and emergency services, and other government agencies including:

- public health department; emergency health investigations;
- EPA–Environmental Impact Aid Department;
- the Federal Emergency Management Agency (FEMA)–Emergency Response Aid;
- State National Guard–Additional manpower and security; and
- other adjacent water supply agencies.

Step 5: Communicate with the Media

At the onset of an alert, emergency condition, or State of Emergency, the designated media spokesperson should be notified. This assignment should be given to a person who can communicate effectively with the media. A media center should be established for both written and verbal press releases. It is also important to monitor media coverage. Sensible public information and communications during an emergency are crucial to the implementation of any type of response plan.

Step 6: Consider Human Health

If an alert, emergency condition, or State of Emergency has been called, it is important to identify the areas most likely to be affected. If health concerns exist, local government officials, local water departments, hospitals, health departments, emergency medical teams, and the fire department must be notified immediately.   Depending on the type and severity of the situation, Federal agencies such as the Department of Health and Human Services, FEMA, EPA, and the Centers for Disease Control and Prevention (CDC) also may have to be alerted.

Step 7: Determine Alternative Sources for Emergency Water Supplies

An emergency situation may require the use of water from other sources, such as unaffected reservoirs, cross-connections to other unaffected water systems or independent water supplies.

[Sidebar: The term "alternative sources of water supplies" as used in this step has a different connotation than alternative water supplies used in both Volume I and Volume II which pertain to water supplies other than municipal water supplies.]

It is important to identify the available sources of emergency water supplies before the need arises. A plan for emergency water distribution and for conveying the water to the consumers should be developed. It is also important to set priorities for where the emergency water will be distributed and how it will be used.

Utilities may choose to partially neutralize the risk of a break in service through a number of avenues that include, but are not limited to:

🔹 Partitioning the distribution system so that certain areas can be shut down without affecting a large proportion of the entire system.

🔹 Have recycling storage tanks of potable water that can be isolated immediately in the event of an emergency; the potable water would be dispensed at the storage tank facility on demand.

🔹 Establish an emergency fire service mobile water tanker program for providing emergency water supplies for fire protection and to support portable water purifying units obtainable from the National Guard, U.S. Coast Guard, or U.S. Navy.

🔹 Estimate hospital and sensitive-user water needs and ensure that potable water would be available to these users in the event of a municipal water system emergency.

Step 8: Establish Relationships with Nearby Water Utilities and Supporting Utilities

Maintaining relationships with other water supplies will expedite a utility's ability to tap into alternative sources of supply in an emergency. Such resources may include independent or private water utilities and purveyors, bottling companies, and some large water-using industries that have their own supplies.

Step 9: Plan and Implement Countermeasures and Recovery Measures

Determine the anticipated types and extent of laboratory tests and establish a testing plan.

Step 10: Supplemental Information Needed for the Response Plan

🔹 formulate an extensive list of the area laboratories, with phone numbers and individual contact names;
🔹 formulate a list of contacts with appropriate State and local government agencies and other companies such as power utilities and bottling companies;
🔹 annually review and update the response plan; and
🔹 maintain evaluation forms and require proper completion if contamination has occurred.

Threat Management

Threat management is a major part of preventing an emergency situation. In the event of a threat, either written or oral, it is necessary to record the contents of the threat and quickly notify the Command Group to evaluate the threat. Each water utility should have an evaluation form that should be filled out by the employee who receives the threatening call or letter.

CRISIS COMMUNICATION

In times of extreme crisis, such as September 11, 2001, citizens appear more trusting of government (and, by extension, utilities). They look to mass media, particularly television, to bring them vital information about how to behave and protect themselves. Therefore, it is critical for utilities to have an effective, efficient plan in place to guide communications with the media, and hence with their public, in the event of crisis. Utilities should not rely on politicians to communicate crises but should keep State, local, and Federal agencies continually updated on the crisis.

It is critical for the utility to communicate in a way that fosters trust and confidence. The utility must be first to release information about how the crisis has affected the drinking water supply. Delayed release of information can result in losses of life, accusation of cover-up, and ultimate loss of public trust. In a crisis situation, the guiding principle for utilities is, "be the first to deliver the bad news." Effective crisis communication has no room for images and egos; it is only about saving lives and protecting the public health.

While the crisis communications plan is being developed, its proposed content and direction should be reviewed with the public through public meetings, discussion groups, civic club meetings, public libraries, and actual tours of water plant facilities to confirm that it meets specific public needs for information. Each segment of the utility's public will have its own needs for information and information delivery, and every effort should be made to accommodate these special needs. For example, the frail elderly may have one set of crisis concerns, pregnant women another, infants and young people another. The utility should develop a list of sensitive stakeholders, contacts, and phone numbers based on the public review of findings. This list must be kept current.

One beneficial outcome of the public review is building a constituency of credible sources with which to partner in case of crises. Potential candidates for credible sources include local universities, the League of Women Voters, parent-teacher organizations, the medical profession, the superintendent of schools, and environmental groups.

PREPARING FOR A CRISIS

The following steps should be taken to prepare for a crisis:

1)   Develop a communication policy and a plan of action for use in case of a crisis.

2)   Form a **crisis team** and define member roles and responsibilities. The team should include representatives from top management, operations, public affairs, government affairs, legal, insurance, human resources, finance, and others.

3)   Compile a list of contact members (email might not be available in a crisis) for each team member and all top management members.

4)   Identify primary and backup spokespersons for the organization. Spokespersons should be authoritative without appearing arrogant.

5)   Notify all utility personnel that only the designated spokesperson should speak with the media.

6)   If the utility uses an outside public relations firm, define its roles and functions. If the utility does not maintain an inhouse staff of public communicators, prearrangement should be made to use an outside firm in the event of a crisis. Should crisis occur, the utility's public relations staff will be very busy. Staff members must be prepared ahead of time with such basic tools as scripts covering various crisis scenarios, lists of frequently asked questions, and media kits, including background information on all possible crisis threats (such as various biological elements).

7)   Develop media (television, newspaper, and radio), community and government contact information (phone, fax, names of contacts), and keep current.

8)   Provide frequent updates to employees regarding the crisis event and the utility's responses and press releases. Post the list of phone numbers for crisis team members.

9)   Request that all employees refer media requests to the designated spokespersons.

10) Establish a schedule of crisis team meetings (morning and afternoon) to update information and issue response to media and employees.

11) Provide media with background information, including information about how the utility protects public health and safety.

12) Request support from the AWWA and other appropriate Federal, State, and local organizations.

13) Work with previously established partners from the medical community to issue joint statements and press releases. Make joint television appearances. Some general guidelines for dealing with media:

   a.   The spokesperson must be in control of his or her emotions, remain calm, appear authoritative but not arrogant, and extremely polite.

   b.   Anticipate likely questions. Understand the perceptions/fears of the public. Address fears by offering facts, not conjectures. Do not answer "what if" questions. Do not use emotional statements or industrial jargon.

   c.   It may be necessary to communicate with the public before all the facts are known. Be humble. State priorities (e.g., getting the situation under control) and assure the audience that the utility staff is doing everything it can.

   d.   Assume that there is no such thing as off the record.

   e.   Avoid conjecture or assigning blame. Both could result in rumors, embarrassment for the utility, and future litigation against the utility.

   f.   Keep communication succinct so as not to confuse or dilute the main message.

SAMPLE THREAT EVALUATION FORM

Type of threat indicated by caller: _____

Specific details of threat: _____
_____
_____
_____

Caller's sex (circle one)   Male   Female   Not sure

Caller's age (circle one)   Under 10   10 – 20   Over 20   Not sure

Describe the voice of the person placing the threat: _____
_____
_____
_____

Describe any background noises or unusual sounds during the phone call: _____
_____
_____
_____

Did the caller name any organization? _____
_____
_____
_____

Did the caller give any other information? _____
_____
_____
_____

Name of the person completing this form: _____

Date and time threat was received: _____

Evaluation form based on Rycus, Snyder, and Meier (1989), Threat Evaluation Form.

Security Analysis & Response for Water Utilities.

References:

1. American Water Works Association. *Emergency Planning For Water Utility Management*. 3rd Ed., Manual of Water Supply Practices–M-19, Denver: Author, 1994.

2. _____. *Distribution System Requirements for Fire Protection*. 3rd Ed., Manual of Water Supply Practices–M-31. Denver: Author, 1998.

3. _____. *Security Analysis & Response For Water Utilities*. Prepared by Nicholas L. Burns, Christine A. Cooper, David A. Dobbins, Jessica C. Edward, and Les L. Lampe of Black & Veatch. Denver: Author, 2002.

Special Sources Consulted:

1. Denileon, G.P. "Introduction. The Who, What, Why and How of Counter-Terrorism Issues." *Journal AWWA* (May), Vol. 93, No. 5: 78-85.

2. Department of Defense. "Introduction." *Department of Defense Chemical and Biological Defense Program; Annual Report to Congress and Performance Plan*: 5-8 (Accessed October 25, 2001).

3. Rycus, M.J., J.C. Snyder, and P.G. Meier. "Strategic Security Planning for Public Drinking Water." Prepared for USEPA, Office of Research and Development. Ann Arbor: The University of Michigan Studies in Urban Security Group, College of Architecture and Urban Planning.

4. Smithson, A.E., and L.A. Levy. "Grounding the Threat of Reality." *Ataxia: The Chemical and Biological Terrorism Threat and the U.S. Response*. Stimson Center Publications. (Accessed November 20, 2001).

5. IS Environmental Protection Agency. *Secrets to Successful SERCs*. EPA 550-F-93-002: 5, 1993.

CHAPTER 9: WATER SUPPLY DISTRIBUTION SYSTEM COMPUTER MODELING

OVERVIEW

Water utilities seek to provide customers with a reliable, continuous supply of high-quality water while minimizing costs. This water often is delivered through very large and complex distribution systems consisting of many miles of pipe and often containing numerous pumps, regulating valves, and storage reservoirs. These systems often are difficult to understand because of their physical complexity, and because of the large amount of data that must be processed. Sometimes, key pieces of information needed to understand a system are not available. In addition, the chemical interactions that take place in the water, and between the water and pipes or reservoirs, are complex. One tool that has evolved over time to help water system designers, operators, and managers in their task of delivering the highest quality, reliable water at a low operating cost is **water distribution system modeling.**

Water distribution system modeling involves using a computer model of a water distribution system to predict the behavior of this system to solve a wide variety of design, operational, and water quality problems. The computer model is used to predict pressures and flows in a water distribution system to evaluate a design and to compare system performance against design standards. The model is used in operational studies to solve problems, such as evaluating water storage capacity, investigating control schemes, and finding ways to deliver water under difficult operating demand scenarios such as a major fire in the community or city. Water quality models are used to perform such tasks as computing water age in the pipe network, tracking chorine residuals, and reducing disinfection byproducts (DBPs) in a distribution system.

Distribution system modeling began with the advent of analog computers and has evolved over time as computer software and hardware developed to become more powerful and easier to use by both engineers and the staff of municipal water system providers. Models containing thousands of pipes are created and used on readily available personal computers. Models that once took hours to run are now run in seconds or even fractions of a second. Originally, models were used only to solve for pressures and flows in the water distribution system. Although this capability remains at the very core of all water distribution system modeling work, hydraulic models or their associated modules now can calculate water age, energy costs, perform isolation analyses, and do optimization of various modeling parameters.

Historically, model building was a very expensive and labor-intensive task. Now that the models can share data effectively with Geographic Information Systems (GISs), Supervisory Control and Data Acquisition system (SCADA), and Customer Information Systems (CISs), the effort to create and maintain a model is reduced. Information obtained from a model study is filtered, organized, and presented in a variety of graphical and nongraphical ways so that the results of a study are more easily understood by a nonspecialist. These advances in technology have broadened the uses of distribution system modeling from just an infrastructure-planning tool to a system used to improve operations, analyze water quality throughout a distribution system, and to plan water system security improvements.

The fundamentals of water system computer modeling as presented in this chapter were, and continue to be, developed by the Engineering Computer Applications Committee of the American Water Works Association (AWWA). The purpose of this chapter is to share an overview of the collective expertise on water distribution system modeling and to present one example of such modeling so that it is understood by municipal officials, water superintendents, and senior fire service officials.

FUNDAMENTAL HISTORICAL DEVELOPMENT OF WATER SYSTEM DISTRIBUTION SYSTEM MODELING

Pre-1970s

Manual calculations or small-pipe systems were used through the 1960s. The Hardy-Cross method was sufficient for single-loop systems, but without the aid of a computer it was impractical for a system having several loops. An analog computer model was created using electronic circuitry in the late 1950s. These were large, physical models that were expensive and difficult to use. Mathematical computer models appeared late in the 1960s. These were simple models that used system data files and system solution algorithms to solve for pressure and flow in the water system.

1970s Through 1980s

Software packages were sold with a variety of features. Steady-state and extended-period simulation models became standard features. Graphics were used for drawing the system and displaying output. Software was appended with the term **packages** because it contained several modular components that were compatible. Some software packages used other specialized software for data entry, display, and reporting of results.

1990s

The 1990s experienced exponential growth of system modeling capabilities. EPANET was a modeling program developed by the U.S. Environmental Protection Agency (EPA) to support ongoing research, and was made available to the public. Some vendors have taken the EPANET model and added an improved user interface. Software packages were designed to be compatible with other standard software packages, such as Microsoft Word®, Excel®, AutoCAD®, databases, and GIS software for editing, database, and drawing functions. The result was a familiar user interface and the ability to use existing software rather than having to create and update new software. Water quality extended-period simulation became standard features in system modeling software packages.

Present

Software packages are working more effectively with, and sometimes as part of, GIS software in response to the dependency on GIS systems by water utilities. GIS data are becoming more common, and the quality of the GIS data is improving, significantly reducing the effort required to develop models. Optimization tools are available for use in optimizing water system design, as well as aiding in the calibration process. Recent security concerns have resulted in studies to develop emergency response plans and to evaluate the impact that various disasters may have on water distribution systems. Models are used increasingly for water quality analysis, such as evaluating water age and DBPs. The EPA also has allowed hydraulic modeling as a means of determining preferred locations for water quality monitoring sites that are required to meet regulatory requirements.

DISTRIBUTION SYSTEM MODELING APPLICATIONS

Benefits of Computer Modeling

Before the use of computerized models, water distribution system analysis using hand-held calculators, conventional slide rules, and Hardy-Cross slide rules involved many simplifying assumptions and approximations. As a result, designs often were much more conservative and expensive than necessary. This implies that larger size pipe was installed than probably was needed, which is contributing to the water-aging problems in older water systems at the present time.

To solve hydraulic system problems, there must be one equation for each pipe section, pump, and valve for each junction point. All of this is dependent on the method used to solve for the unknowns in the hydraulic calculations. The number of equations that must be set up and solved in a system hydraulics problem is very large, even for the most basic water distribution system. The value of a computer model is that tedious calculations are performed very fast and more accurately than by manual calculation. In addition, the computer is effective in managing the large amounts of data necessary to analyze a water distribution system. By using computer models, decisionmakers can focus on formulating and comparing alternatives as well as communicating results, rather than on the procedural mechanics of solving system equations.

Computer models of water distribution systems are not an end in themselves but are a tool to help managers, engineers, planners, and operations staff. Their real purpose is to support decisionmaking processes in planning, design, and operation of water distribution systems. When properly implemented, the model is an integral part of the utility's decisionmaking process. Ultimately, engineers and operators of a water system still are responsible for decisions based on output from computer models.

Distribution system modeling software generally falls into four application categories: planning, engineering design, systems operations, and water quality improvement.

Planning

A primary planning application of distribution system analysis software is in assisting development of long-range capital improvement plans, which include scheduling, staging, sizing, and establishing primary routing and location of future facilities. Other applications include the development of main rehabilitation plans and system improvement plans. Rehabilitation plans identify and prioritize mains that need to be cleaned and/or relined. Distribution system improvement plans identify where the installation of new water mains, storage facilities, and pumping stations are necessary to keep pace with growth or new utility standards. The following are examples of specific system analysis planning applications:

- Capital improvement programs. Water utilities usually have a master plan that identifies capital improvements. A model usually is used to identify these capital improvements to respond to projected growth or to replace aging infrastructure.

- Conservation (impact of) studies. Water conservation is used to stretch limited water supplies or to reduce water use so that capital improvements are delayed or eliminated. A model is useful to apply projected demands with conservation measures to evaluate their potential for success.

- Main rehabilitation program. A model is used to identify the effect of specific mains that are bottlenecks to the system, either because demands have increased significantly or because of encrustation. The model is used to determine the hydraulic effect of rehabilitation alternatives.

- Reservoir siting. A reservoir should be located where there is good turnover in the reservoir, where the reservoir effectively meets peak demands, and is filled during off-peak demand times. The model is used to explore these scenarios to fine-tune preferred hydraulic solutions.

Engineering Design

Engineering design applications include the sizing of various types of facilities. Pipelines, pump stations, pressure-regulating valves, tanks, and reservoirs are sized using pressures and flows that result from distribution system modeling. In addition, system performance is analyzed under fire-flow conditions and adjustments are made to meet fire demand. Following are examples of engineering design problems that are solved using computer models.

- Fire-flow studies. The model is used to simulate fire-flow demands at hydrant locations throughout a community (city) to determine how much water is delivered at one or more flowing hydrants with a minimum residual pressure of 20 pounds per square inch (psi) at each hydrant and maintaining a minimum of 20 psi residual pressure at representative locations on the water distribution system. When deficiencies are discovered, the distribution system needs to be reinforced by one of several methods, including pipe looping. These studies also are used to demonstrate compliance with insurance public protection criteria and building code requirements.

● Valve sizing. A distribution system often has pressure-regulating or pressure-sustaining valves to direct flow to a different demand zone. Distribution systems also may have throttle valves to direct flow within a specific zone to different reservoirs or storage locations. The model is used to determine how much flow is required through these valves, so that the valves are sized appropriately.

● Reservoir sizing. Reservoirs often are sized by estimating the total diurnal flow, fire flow, and emergency storage requirements within a particular zone. However, reservoir capacity also should consider the rate of water delivery to the reservoir location and the size of the distribution area. A model is useful to evaluate inflows and outflows to a reservoir in order to determine an optimal size for a particular location or to specify other improvements so that a reservoir at the preferred site is served adequately by transmission mains and pumping stations.

● Pump station/Pump sizing. Models are used to calculate system curves of distribution systems so that pumps are selected that provide the necessary head and flow. Proposed pumps then are used in the model across the range of operating conditions to determine how well they meet a variety of operating conditions.

● Calculation of pressure and flow at particular locations. A water supply distribution system must provide adequate amounts of water at pressures that are within a range specified by the standards used by the water utility. A model's core functionality is pressure and flow hydraulic calculations. Models are used to predict pressures under specific conditions and under a wide variety of scenarios to identify low pressures and to select the infrastructure that will remove flow or pressure deficiencies.

● Zone boundary selection. Most water distribution systems deliver water to customers located at a range of different elevations. Distribution systems are separated into pressure zones that follow consistent elevation contours in order to determine zone boundaries and the adequacy of the infrastructure that delivers water to each zone.

System Operations

Applications of operations include assisting in the development of operating strategies, operator training programs, and system troubleshooting guidelines. Emergency conditions, energy management, water availability, etc., may drive operating strategies. For example, contingency plans are developed in the event a key facility, such as a pump station, fails. Distribution system modeling also is used to develop operational strategies for energy management and water quality guidelines. Strategies for shifting supply among treatment plants are developed that determine efficient use of available water, optimizing these strategic results in efficient use of pipeline capacities, water tank levels, and required treatment plant production, among other things.

💧 Personal training. Models are used for training personnel who operate the water distribution system. System operators can experiment with the model to determine how the system responds to changes in operating conditions.

💧 Troubleshooting. Models are used to troubleshoot the cause of various problems, such as low water pressure, water circulation problems, and events that otherwise would be inexplicable.

💧 Water loss calculations. In the event of a major water main break, the model is used to estimate the amount of water lost through the break, as may be required for damage assessment.

💧 Emergency operations scenarios. Water distribution systems often have critical components; if the components fail, water delivery is interrupted. A model is useful to evaluate the potential impact of a failure and to devise means of reducing the damage or impact of a critical component failure.

💧 Load shifting between treatment plant studies. Water treatment plants are sometimes taken out of service for repairs or because the water supply is unavailable at the time. Furthermore, the quality of water at one source may be better at certain times of the year, so the use of high-quality water sources is maximized. The model is useful to devise operating scenarios to use water sources to achieve desired objectives.

💧 Model calibration. Model calibration typically is thought of as a step in developing a useful model. However, the calibration process is useful to operations staff in discovering anomalies in the distribution system, such as closed valves, tuberculated pipes, pipe leaks, or incorrect infrastructure data. This information, once discovered through the calibration process, can explain operational difficulties, and identify distribution system problems that will improve the operation of the system when resolved.

💧 Main flushing program. A hydraulic model is an excellent tool to develop a main flushing program. A model is useful to identify flow paths in the distribution system so that the flushing locations and sequence can be established.

💧 Area isolation. Water utilities frequently need to isolate an area for maintenance or other work. Often, it is helpful to identify those customers whose service will be interrupted by the isolation event. In addition, those planning the event need to know which valves to close in order to minimize the impact of the isolation. Hydraulic modeling is used to improve water quality.

Water Quality Improvements

Water quality regulations in the United States are regulating the level of DBPs in a water distribution system. Standards and expectations for water quality have increased, as has demand for water quality analysis in the distribution system as reflected in Volume I. Following are examples of how distribution system modeling is used to improve water quality.

◉ Substance tracking. If a contaminant enters the distribution system through a treatment plant, well, reservoir, or other location, it will spread throughout the distribution system, affecting the water quality of consumers who receive water from that source. The contaminated water also may mix with water from other sources. A model is useful to predict the contaminant level and zone of influence of the contaminant in the distribution system. Customers affected by the contaminant are identified, and portions of the distribution system that need to be flushed also are defined.

◉ Water source/Age tracking. It needs to be stressed that water age is an important water quality parameter in a distribution system. Chlorine levels decay over time, and DBP levels tend to increase with time as the chlorine reacts with organic compounds in the water. To maintain water quality, water utilities are striving to minimize water age. This is done by ensuring that water in reservoirs turns over regularly, so that it does not become stagnant, and by minimizing dead ends. When multiple water sources serve an area, operating strategies are devised to reduce water age where possible. Distribution system modeling helps identify operating strategies to reduce water age.

◉ Chlorine levels. A model is used to predict chlorine decay in a distribution system. This is useful to determine chlorination levels at the treatment plant and to select rechlorination sites where necessary to boost chlorine levels.

◉ Water quality monitoring locations. Part of the water quality regulations developed by the EPA include selecting appropriate sites to place permanent water quality monitors to monitor DBP levels and demonstrate compliance with Federal regulations.

HYDRAULIC MODELS

A computer model is composed of two parts: a database and a computer program. The database contains information that describes the infrastructure, demands, and operational characteristics of the system. The computer program solves a set of energy, continuity, transport, or optimization equations to solve for pressure flows, tank levels, level position, individual pump status, water age, or water chemical concentrations. The computer program also aids in creating and maintaining the database, and presents model results in graphical and tabular forms.

Model Data

Model data usually is associated with two entities: links and nodes. Links represent pipes, pumps, valves, and sometimes tanks in a distribution system. Nodes represent junctions, endpoints, and locations of elevation extremes, water sources, demand points, and both ground-level and elevated water tanks. The characteristics of facilities, such as pipes, include length, roughness or friction coefficient, diameter, and reaction coefficient. Operational parameters also are associated with the respective links. Attributes of nodes include elevation, grade line, water demand, and water supply. Model databases include meta-data and descriptive information that is useful in defining, organizing, and managing the model.

The model is a valuable asset to the user, and represents a substantial effort in data collection, entry, and quality control. Therefore, this investment in data and in an understanding of the distribution system should be protected. One of the keys to maintaining the value of a hydraulic model is **making sure** that the model is updated with new or confirmed data elements, infrastructure, demands, and operating instructions. One of the most critical numerical values of the model data is the calculated "c value" or coefficient of roughness for the pipe wall of each pipe section. Guesswork on "c values" is not acceptable, and can lead to erroneous output information.

In order to maintain confidence in the results of the model, data within the model should represent the current system configuration so the stakeholders will trust and use the model.

Modeling Software

The heart of water distribution system modeling software is a system solution algorithm, which sets up and solves the hydraulic and transport equations. Depending on the problem, several kinds of equations are solved:

- Continuity equations keep track of water flow, make certain that the total flow into the node equals the flow out plus or minus any changes in storage or demand.

- Energy equations account for energy loss caused by friction in pipes, valves, and fittings, and energy gained in pumps (i.e., the total energy change around a pipe loop is **zero**).

- Transport equations are required in water utility models to account for the movement of substances (pollutants or tracers) through a distribution system and any reactions that may occur.

- Cost equations are needed in optimization models to account for energy costs or cost of piping. In addition to a computation engine, the software must manage the databases or files containing model information, interact with the user, and present model results in both graphical and nongraphical forms. The following features are available in distribution system modeling software packages, and should be included in any modeling package for use by a water utility.

Steady-state analyses. Packages will have the ability to perform a steady-state analysis, which takes a snapshot of pipe conditions at any instant in time. Steady-state analyses typically are used to evaluate maximum day, peak hour, and fire-flow conditions.

Extended-period simulation. A package also should have the ability to perform a sequence of analyses, with the output from each analysis forming the input to the next analysis. This capability may be used, for instance, to model the operation of a given water system over a 24-hour period with an analysis run each hour. Such a simulation is useful in modeling variations in demand, reservoir operations, water quality, and water transfers through transmission pipelines. Extended-period simulation requires that the system package model flow and pressure switches incorporate demand hydrographs for nodes, and allow for varying water tank configurations.

Graphical user interface (GUI). A package should provide a graphical interface, which allows the user to see a schematic of the water distribution system displayed on the monitor screen. Link and node data are displayed by clicking on the entity on the screen, and model results also are displayed for each entity. Modeling results are presented graphically and in tabular form, giving the modeler a better understanding of distribution of flows and pressures throughout the system. GUIs can be either proprietary (written specifically for the package) or customized and generated from standard graphics packages, including Computer-Aided Design and Drafting (CADD or CAD) packages.

Error reporting. A package should test for distribution system configuration errors. For example, any portions of the distribution system that are not connected to the rest of the system should result in an error message indicating the location where the model is discontinuous.

Selective reporting of results. The user is able to specify the results to be reported in tabular form, so pages of output need not be generated after each run, as the user may be interested only in results in one specific area of the water system. This user-specific reporting saves hard-disk space and paper while speeding the user's review time.

System components. As well as pipes and nodes, a package should have the ability to model pumping units (incorporating pump characteristics curves) as well as flow and pressure-regulating devices.

Data management. Modelers should be able to export and import data and model results to and from other applications, such as spreadsheets, databases, and GIS systems. These capabilities are widely available, and an important part of any modeling package.

Automated fire-flow calculations. Some more advanced distribution system modeling packages can calculate the available fire flow automatically at each node or fire hydrant on a real-time basis. This information can be relayed to the first-responding engine company for the fire hydrant designated as the initial-supply fire hydrant for a specific risk site. This should be a significant advance in the effective confinement and extinguishment of developing structural fires. These calculations also are useful for identifying areas on the distribution system where the available fire flow does not meet the Needed Fire Flow (NFF).

Scenario generation. Distribution systems with any level of complexity are modeled more easily by applying various combinations of demands, facilities, and operating parameters (such as regulating valve and pumping unit settings). System modeling packages may allow variation and combination of these three types of data in a simulation by keeping them in separate databases for specific or combined model access.

Water quality. Utilities increasingly are interested in modeling the water quality within a distribution system, particularly the decay of chlorine residual and water age. The ability to perform water quality analysis should be a standard part of any modeling package.

Hardware

Typical hardware for modeling consists of a personal computer, printer, and plotter. The computer should meet the modeling software requirements. Color plotting or printing is helpful to communicate model results. If the organization is large enough to have a computer network and a network server, the modeling system should use the server for data backups and storage. Regular data backups are required. The software vendor should be consulted for specific hardware recommendations at the time software is purchased.

Related Software Systems

Information management trends within utilities are moving toward better information sharing so that decisionmakers can have many information resources to make the best decisions. Often this is done by using both common databases and files that are shared by a variety of information about physical assets, customers, billing data, geographical information, and operational information. Furthermore, modeling activities can benefit a variety of groups within the utility, strengthening the need to communicate and share information. Brief descriptions of software systems, information systems, or corporate-wide databases that are in some way related to distribution system modeling are listed as follows:

- Geographical Information System. A GIS stores and displays information that is best understood by displaying on maps. The spatial relationships between entities are significant for most information that is stored in a GIS. A GIS has the potential to store vast amounts of information that are useful for systems analysis. These data can include pipe assets, customer meter locations, zoning and land parcel data, aerial photographs or other land bases, street locations, digital terrain models (DTMs), and jurisdictional boundaries. Information in a GIS is saved in file formats that some modeling software packages can read. Alternatively, information in a GIS database is translated into a form that is imported into the model database. Some GIS land-base information is collected and stored in the GIS database. Pipe data in the GIS is most useful if the topology, or connectivity, is already established in the GIS.

- Computer-Aided Design and Drafting. CADD systems are used to create maps and of water supply distribution systems. Therefore, they are a source of pipe information that can be transferred to the model. In addition, CADD systems are a useful means of displaying model information and results. Some modeling packages bring CADD information into the model using Drawing Exchange Files (DXFs).

● Supervisory Control and Data Acquisition. SCADA systems are used to control the operation of pump stations, valves, and other system infrastructures remotely. They also are useful to collect data that includes pressures, flows, reservoir levels, valve positions, pump status and speed, chlorine levels, and other information that is useful to monitor the system. This information is collected at frequent intervals and stored for extended periods of time. SCADA is a good source of operational information, as well as calibration data. SCADA data also are used to define the boundary conditions that are placed in the model. SCADA data usually do not go to the model directly. An interface usually is required that could be as complex as a custom software routine, or as simple as importing SCADA data into a spreadsheet via a CSV file and formatting the data to import it into the model.

● Customer Information System. A CIS is useful to provide customer water-use information to develop demands. Typically, average annual water usage and customer rate class for each customer is extracted from a CIS; then GIS modeling or customized tools are used to link these demands to nodes in the model. The specifics of how the CIS data are linked and entered into the model is highly dependent on the data and software used at the utility.

DISTRIBUTION SYSTEM MODELING WITHIN THE UTILITY

A successful distribution system modeling program functions best with a team of individuals who fill all the roles required for a system analysis program and are able to provide system modeling results effectively to decisionmakers within the utility. Issues that often need to be addressed when implementing a modeling program are outlined as follows.

In-House Versus Outside Consultants

The utility should decide whether the model is developed and maintain by the utility or by an outside consulting firm. Usually, a utility understand its system very well and has easy access to model-related information. However, a utility may not have the expertise or resources to develop and maintain the model. Some utilities construct and run their own models, while others hire outside engineers and consulting firms to perform some or all of the modeling work. A model owner who is committed to maintaining the model and to developing modeling expertise is essential for an in-house modeling program. If consultants do the modeling, ownership rights to the model should be delineated clearly in a contract. Regardless of who does the modeling work, a long-term commitment to maintaining the model and to having experts use the model regularly are necessary components for success.

One-Time Versus Long-Term Use

Many decisions made during model development depend on whether the model is used for solving a short-lived problem or for periodic use over an extended period of time. If the model is used for a specific problem, questions regarding the level of detail are answered easily, based on satisfying the needs of the problem. If the model is to be used for many purposes, the model should be developed to serve the most demanding applications and simplified, if necessary, for other applications. For example, a model developed to assist in master planning may not contain enough detail for determining available fire flows in subdivisions or for water quality modeling. Decisions must be made on whether this level of detail should be included in the model or added later, if required. The power of today's computers makes it practical to use a complex model for even a simple task. However, many software packages have tools to simplify a model when necessary.

Model Developer Versus Decisionmaker

There are two distinct roles in model development: the role of the modeler and the role of the decisionmaker. The modeler is a person responsible for initial development and running of the model. The decisionmaker is a professional engineer or licensed operator who interprets and makes decisions based on outputs from the model. These two roles could be filled by one individual, two individuals, or even two different companies. The key element is that the decisionmaker must be satisfied that the modeler has indeed developed a system model that is adequate for the problem or problems being considered. Calibration and sensitivity analysis are two methods available to the decisionmaker to ensure that the model is adequate for the intended purpose.

Modelers Versus Rest of the Utility

It is essential that modeling, whether performed in-house or by a consultant, be done with an awareness and cooperation of the rest of the utility. While some individuals may serve as the experts on the model, all interested parties should have input in model development, understand the capabilities and limitations of the model, and appreciate the important role of modeling in decisionmaking. Modeling should be done with through consideration of utility operations. For example, utility operators have great insight into the operation of a system, as well as physical limitations. By working with the operations staff, the modeler can incorporate operator's insights into the model, and operations staff can become sufficiently comfortable with the model to trust its results.

SURVEYS

Since 1980, the AWWA Engineering Computer Appellations Committee and other groups have surveyed water utilities to compile information about the use of system modeling packages. In the 1999 survey, a total of 989 utilities were selected from the AWWA database as serving greater than 35,000 people. A total of 174 water utilities responded to the survey, 150 of these utilities were using a distribution system modeling package. The paper *Hydraulic and Water Quality Modeling and Distribution Systems: What Are the Trends in the U.S. and Canada?* discusses the results of the survey. (SR-1) Notable results of the survey follow:

- A total of 20 different distribution system modeling packages were used by the respondents, but the market was dominated by a relatively small number of packages. System modeling packages are typically used weekly or monthly.

- Seventy-seven percent of respondents described the accuracy of their model (accuracy of 60 to 100 percent); 97 percent gave no response at all.

♦ Most respondents used their model for steady-state analysis with only a few performing extended-time simulations.

♦ Billing records typically were used to generate node demands.

♦ Few utilities linked their modeling packages to other software, but many had plans for multiple links in the future.

TRENDS

Several significant trends in water system modeling have become apparent through the 1999 Network Modeling Survey, presentation and discussions at conferences, and *Journal AWWA* articles. (See Special References.) Some trends now are well-established, while others are still in their infancy.

♦ Use of personal computers: Personal computers (PCs) now comprise the dominant hardware used for system modeling. Some large utilities maintain models on network workstations, allowing simultaneous access to a centralized database. However, PC-based system modeling packages easily model water distribution systems containing thousands of pipes in seconds.

♦ Graphical user interfaces: Distribution system modeling packages are capable of graphical input and output, easing data entry and evaluation of results.

♦ Water quality analysis: Many system modeling packages have the ability to model water quality parameters in a pipe network and in reservoirs, which utilities find valuable in response to new water quality regulations and to heightened public awareness of water quality.

♦ Common databases: In some utilities, computer systems are organized around a type of architecture where common databases are shared by many applications. In such a framework, a distribution system modeling package extracts data from a large enterprise database that is shared by many work groups.

♦ Model sophistication: Model surveys reveal a strong trend toward all-main models and extended-period simulation to examine water quality issues in water distribution systems.

♦ Demand allocation: Although no new demand methods were identified in the latest survey, there was greater emphasis on using GIS tools in model loading.

♦ Frequency of use: Utilities intend to use water models more frequently in the future. Specific uses include mater planning, fire-flow evaluation, and energy and operations optimization.

♦ Information systems integration: The near term will see a far greater integration of information systems such as GIS, SCADA, and CIS.

♦ Real-time modeling: In the past, water distribution system modeling packages were typically too slow and unwieldy for system operators to use in generating operating strategies and testing what-if scenarios. High-speed processing and data input available from SCADA allow utilities to provide modeling capabilities to their operators. Careful consideration must be given to the user interface in this regard, and simplified models may be required.

Primary Reference:

American Water Works Association. *Computer Modeling of Water Distribution Systems*, Manual of Water Supply Practices–M-32. American Water Works Association, 6666 West Quincy Avenue, Denver, Co. 80235-3098.

Special References:

1. Anderson, Jerry L., Mark V. Lowry, and James C. Thomte. "Hydraulics and Water Quality Modeling of Distribution Systems: What are the Trends in the U.S. and Canada?" Denver: *Proceeding of the AWWA Annual Conference*, 2001.

2. Bhave, P.R. *Analysis of Flow in Water Distribution Systems*. Lancaster: Technomics Publ., 1991.

3. Eggener, C.L., and L. Polkowski. "Network Modeling and the Impact of Modeling Assumptions," *Journal AWWA* 68:4 1996, 189-196.

4. Haestad Methods, Inc. *Computer Applications in Hydraulic Engineering*. Waterbury: Author, 1997.

5. Walski, T.M., J. Gessler, and J.W. Sjostrom. *Water Distribution–Simulations and Sizing*. 2nd Ed. Ann Arbor: Lewis Publishers, 2000.

CHAPTER 10: ESTABLISHING A COMMUNITY PROGRAM TO DOCUMENT EFFECTIVE COMMUNITY WATER SUPPLIES FOR FIRE PROTECTION

WHERE ARE WE NOW?

The first nine chapters of this Volume provided a foundation for understanding the components of a community water delivery system starting with a raw water source and ending with the need to provide an adequate and reliable water system that will meet consumer consumption needs plus fire-flow demand throughout a service area. It is understood that the consumer consumption requirements fluctuate, sometimes significantly, by time of day, day of the week, and month of the year. Regardless of these fluctuating demands, a hostile structure fire may be in progress anywhere in the community and water needs to be available almost instantaneously at one or more fire hydrants to supply fire apparatus to confine, control, and extinguish the fire in order to save lives and minimize property damage. The fact that **working** structure fires are considered a rare event in communities up to 50,000 population results in some challenges in providing the economic base to reinforce an existing water supply system to provide maximum daily consumption flow rates **plus** adopted building code water requirements for fire protection and/or Needed Fire Flows (NFFs) established by the Insurance Services Office, Inc. (ISO).

In order to establish a community program to provide adequate and reliable water supplies for the future, community official and leaders, water utility leaders, and fire department officers need to answer the question, Where are we now? This translates to understanding the capability of the water supply system to deliver treated water to **all** water supply consumers under conditions of average daily consumption and maximum daily consumption. There also has to be an understanding of the water system's capability to provide NFFs at representative fire risks throughout the community.

GETTING STARTED

An excellent starting point is to review the latest ISO Public Protection Classification Survey report pertaining to the community. It should be recognized that these survey reports are updated by new surveys conducted approximately every 10 years by an ISO Field Representative. The community mayor, city manager, or in some case the county executive has the authority to request Public Protection Classification reports by contacting the regional office of ISO. A listing of the regional offices is provided with the references at the end of this Chapter. (Reference #1, pg. 181)

After reviewing the grading report on water supplies, it may be appropriate to request that the assigned ISO Field Representative meet with the effective water supply committee to discuss any improvements in the water supply system that would have a positive effect on the deficiencies document in the report. During this exchange of information, it is appropriate to conduct current fire-flow tests at the representative fire risks named in the last grading report for comparative purposes. These flow tests are to be conducted in accordance with the procedures established under Chapter 3 of this manual. It is important to conduct these tests when the daily consumption records indicate approximately the same amount of water usage that was identified when the ISO NFF tests were conducted. A graph of the ISO flow tests at each fire risk site should be prepared and then plotted with the results of the current fire-flow tests. This permits a comparison of flows at the same pressures. This concept is depicted in **Figure 10-1**. Curve 1 shows the results of the ISO water flow tests conducted 6 years ago. The Available Fire Flow (AFF) at the subject building, which is a hardware store, was 1,500 gallons per minute (gpm) at 20 pounds per square inch (psi) residual pressure, while the NFF was 1,400 gpm. At this specific fire risk site full credit is given for the water supply. However, a recent flow test at the same location conducted by the water department and the fire department indicates a current available flow of 900 gpm at 20 psi.

Figure 10-1

Comparative Analysis of Water Supply Curves at Dietzen's Hardware Store

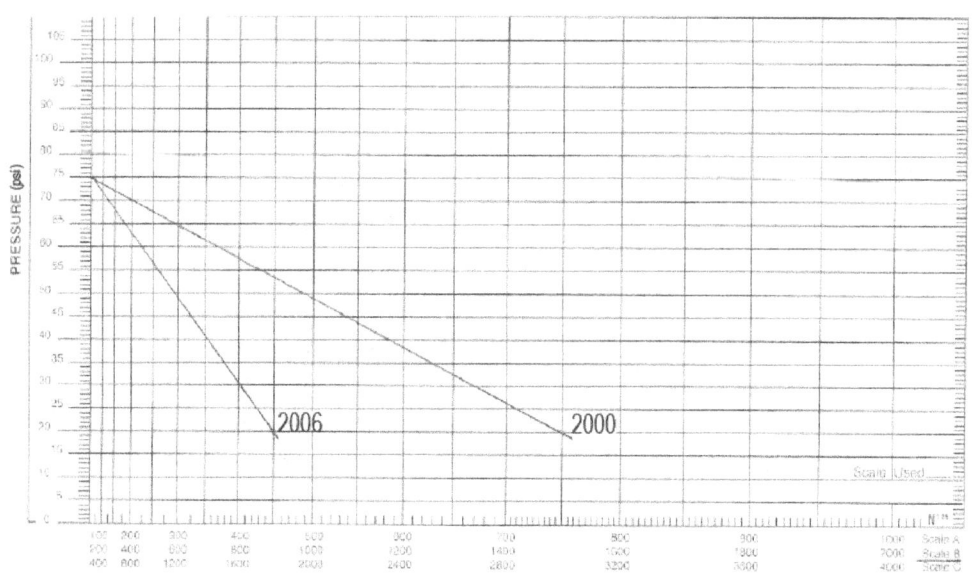

The comparative analysis shows the same referenced static pressure, so consumer consumption in the area is considered the same for both tests. However, the flow rate has diminished by 600 gpm, or a 20-percent drop. A change in flow characteristics in flow tests conducted at the same location and in the same manner that equals or exceed a 10-percent reduction need to be investigated.

Since the credited fire flow at each representative fire risk is determined at 20 psi residual pressure, the comparisons should be evaluated at this common pressure. If the static pressure points are not the same, it is a possible indication that the flow demand in the test area has changed. Lines that are parallel indicate that the characteristics of the pipe network are the same for both tests. For a reason to be determined, the fact is that the flow and pressure entering the distribution system has changed.

COMPARISON TO BE MADE OF THE PHYSICAL FEATURES OF THE WATER SYSTEM

The water supply study group needs to make a comparative analysis of all features of the water system between when the last ISO *Grading Report* was published and current conditions. As a minimum, the following questions need to be answered under each topic heading. This is a general list of issues, and is not to be considered all-inclusive for every water system.

1. Domestic consumption.

    a. What changes, if any, are identified for the following classification of water consumption?
    - Peak consumption hour. [Note: It is not unusual for a water system to have two peak periods in a 24-hour evaluation time.]
    - Average daily consumption.
    - Maximum daily.
    - Monthly maximum consumption.
    b. Can graph recorders identify the amount of water over time used by the fire department to fight structural fires?

2. Raw water source and storage.

    a. What is the source or sources of raw water for the community?
    b. Has there been any change, especially in capacity, of the raw water source or sources?
    c. Is there a raw water storage facility such as a tank or pond that is independent of the primary raw water source for the community?
    d. How does the raw water source feed the water treatment facility, and have there been any changes to this operation?
    e. Has the raw water source been evaluated by a hydrologist or geologist to determine the minimum storage capacity during a 50-year drought and/or freezing cycle?

3. Water treatment facility.

    a. What is the maximum flow rate in gpm or gallons per day (gpd) for converting raw water into finished water?

b. Does treated water go directly to a clear well, a ground-level storage tank, an elevated storage tank, or is it pumped directly into the distribution system?

c. Does the finished water meet current Environmental Protection Agency (EPA) criteria for clean water?

d. Has there been any evidence of contaminated water?

4. Water pipe distribution system.

a. What changes have been made to the water distribution system? The question covers new pipe installations by size and location, pipeline repairs and replacements, the installation of pipe valves and the location of these valves, the installation of new fire hydrants, and the location of each hydrant.

b. Has the coefficient of roughness (i.e., the "c value" in the Hazen-Williams pipe formula) been determined for the primary feeders, the secondary feeders, and the distributor mains?

c. Does the water department or authority have instrumentation for determining minimum and maximum flow velocities in feet per second for pipeline evaluations?

d. Do all fire hydrants meet the specifications and installment methods specified by the American Water Works Association (AWWA), *Manual of Water Supply Practices: Installation, Field Testing, and Maintenance of Fire Hydrants* (M-17)?

5. Water supply component testing and checking.

a. Are all fire hydrants flow tested semiannually and the results recorded on hydraulic graph paper for comparative analysis between test results?

b. Are all installed fire hydrants flushed and maintained annually or at the time of flow tests?

c. Are fire hydrants that have been placed out of service for whatever reason tagged with an out-of-service collar?

d. Are water supplies to commercial properties equipped with automatic sprinkler systems evaluated on a semiannual basis?

e. Are all water system valves maintained and turned down and then out, counting the turns, annually?

6. Professional engineering services.

a. Does the community employ a civil, mechanical, or fire protection engineer who can assist the water department, company, or authority?

b. Does the community retain the services of a professional engineering group to assist the water department, company, or authority?

7. Water distribution map(s) and drawings of facility installations.

a. Does the water department, company, or authority have in place a current community map showing all of the street, underground pipe by size and location, water system valve locations, fire hydrant locations, and water storage facilities?

b. Are detailed layout drawings available of the water treatment facility, pumping stations, and both ground-level and elevated storage facilities?

8. What expansions are planned to meet new community growth, such as housing developments and commercial expansion?

9. Has the community formed a water supply planning committee?

a. The above outline demonstrates that many tasks need to be accomplished to make a comparative analysis between the condition and capabilities of the water system today and when the ISO conducted the last Public Protection Classification Survey.

b. As a minimum there needs to be a small group from the: 1) water department or water authority to represent all phases of involvement with the existing water system; 2) fire department who need to update the NFFs at representative fire risks throughout the community and establish the location of all buildings protected by automatic sprinkler systems along with determining need water supply and riser pressure for each sprinkler system. Fire service personnel should be on a team with water department personnel to conduct fire-flow tests and prepare flow graphs of each test. This is essential information for fire department prefire planning of commercial and industrial properties; and 3) community elected officials and full-time administrators need to be assigned to working groups for understanding the water system review process, because they are going to be important advisors and decisionmakers when it comes to financing needed water supply improvements.

The study committee needs to ask, How does the existing water supply system compare to the water system of 5 or 10 years ago in meeting the water demand for both consumer consumption and NFF? The answer to this question should be in the form of a report prepared by the committee.

EVALUATE OPTIONAL METHODS AND PROGRAMS TO MEET THE COMMUNITY'S CURRENT AND FUTURE WATER SUPPLY DEMANDS

It would be very unusual if the current water system resources and distribution system capabilities meet the needs of any community, especially when looking to future growth of the community. Therefore, the next step is to evaluate an **action program** for meeting current and projected deficiencies in the total water supply system from raw water resources to the delivery of treated water that meets EPA criteria for clean water to consumers that use water for drinking, cooking, and personal hygiene, **plus** water supplies to meet building code-required water supplies and ISO NFFs. One very important option to consider is the installation of automatic sprinkler systems as a program to reduce any further impact on the community water system, to provide fire protection water, and possibly to reduce some of the large fire-flow rates that currently exist. This is the first item presented below.

It needs to be recognized that the water supply for firefighting using firehose streams is not required to meet the same quality of water standards as drinking water. This understanding provides communities with several potential options for improving the water supply for fire protection without interacting with the domestic water system. One or more of these options may be more cost effective and more efficient than trying to upgrade the existing water system to supply both domestic water and water for fire protection. Any water supply study group should keep these options in mind as they evaluate the current condition and capabilities of the extinguishing water supply system.

1. Installation of automatic sprinkler systems: As presented in Volume I, the AWWA Research Foundation and KIWA of the Netherlands combined to produce a published document titled, *Impact of Fire Flow on Distribution System Water Quality, Design and Operation*. The research group of over 50 professionals from all disciplines and experiences associated with municipal water supplies determined that large-diameter piping that was installed to meet both consumer consumption and NFFs resulted in water aging that reduced the quality of water in pipes below EPA and other public-health criteria for safe drinking water. This occurs because of the infrequent use of fire flows over 500 gpm needed to extinguish developing hostile fires, especially in residential areas. This study set forth a number of options for solving this problem. However, at the conclusion of the discussion relating to these options, the study report states,

    > In particular, the universal application of automatic sprinkler systems provides the most proven method of reducing loss of life and property due to fire, while at the same time providing faster response to a developing fire and requiring significantly less water than conventional fire-fighting techniques. It is recommended that the universal application of automatic fire sprinklers be adopted by local jurisdictions. (Reference #2, pg. 182)

    To advance this type of program, a number of communities have offered incentives to commercial and industrial property owners to install automatic sprinklers in new buildings and to retrofit existing buildings with sprinkler protection. Some of these incentives include: (Reference #3, pg. 182)

    a. Property tax relief (examples vary greatly).

    b. Install the underground piping and valves needed to supply domestic water to the sprinkler control valve and install backflow prevention devices that may be required by adopted codes.

    c. Conduct and record the required 2-inch-drain flow tests on a semiannual basis by city employees (i.e., conducted during fire department in-service inspections).

Another important incentive is the reduction in insurance premiums for most classes of commercial property. Case studies show that a wet-pipe or dry-pipe sprinkler system will pay for itself in 10 years or less as a result of lower property, casualty, and liability insurance rates. Furthermore, The ISO *Fire Suppression Rating Schedule* does not calculate a NFF for a sprinklered building; National Fire Protection Association (NFPA) 13, *Standard for the Installation of Sprinkler Systems*, is used to calculate the flow for the sprinkler system and the supplemental hose streams.

It needs to be understood that, until further studies are make on optional programs for providing needed water supplies to a community, potable or treated water is the only acceptable water supply for automatic sprinkler systems.

2. The community water system would provide a 500-gpm fire flow for all property throughout the community: This concept specifies that a given community would supply the minimum required fire flow to all structural property at a residual pressure of 20 psi. Why select 500 gpm as the criteria? The answer is that this is the minimum requirement for water supply to built-upon areas by the ISO. If this flow rate is satisfied, then the question becomes one of identifying all properties with NFFs over 500 gpm. This type of information can be requested, as previous discussed, from the regional office of the ISO. (Reference #1, pg. 182) Now there are four basic approaches to providing the difference in flow rate between 500 gpm and the NFF. For example, a furniture store and warehouse has a NFF of 2,000 gpm, but only 500 gpm is available from the domestic water. How can the other 1,500 gpm be supplied at the fire-risk site, not considering the installation of an automatic sprinkler system as discussed above?

   a. Dual water system: Consider establishing a dual water system that uses reclaimed water from the sanitary system to provide separate piping that would loop around the commercial/industrial areas that need the higher fire flows. The system would be designed to meet the maximum fire-flow differential unless there is a single fire risk that has a very abnormally high NFF. Fire hydrants would be installed as needed to meet the requirements for each specific fire risk. The advantage of this method is that the reclaimed water also can be used for street cleaning, watering grassy areas, trees, and shrubs, plus other uses where potable water is not required.

   b. Separate water system: Separate water systems for fire protection go back to the late 1800s, and two examples of separate water systems for fire protection use only are presented in this manual: one in Baltimore, Maryland, and one in New York City. There are examples of small separate water systems to provide fire protection water just in the central business district of small communities, or to protect a specific fire risk. After a study of needed fire flows in a community in relation to available fire flows, it may be important to evaluate the feasibility of developing a separate water system to meet specific fire protection needs.

   c. Alternative water supplies: Alternative water supply operations typically involve using mobile water tankers to shuttle water from a recognized water supply, such as ponds with 30,000 gallons of water or more, reliable running streams and rivers, and lakes of all sizes to specific fire sites. Fire departments in small and medium-size communities often operate mobile tankers to provide fire protection for structures located outside of 1,000 feet of a fire hydrant. These tankers and other tankers operating under automatic-aid and/or mutual-aid agreements can be used to augment the water supply provided by the community water system to provide or meet NFFs for specific properties. An ISO Field Representative needs to witness a water delivery demonstration to credit this supplemental water supply.

[Sidebar: See the referenced document by the Federal Emergency Management Agency (FEMA) on *Alternative Water Supplies* at the end of this Chapter.]

SUMMARY

One of the best tools for analyzing community water supplies is the preparation of water supply summary sheets as presented above and in Chapter 3. The graphing technique based on the Hazen-Williams Pipe Flow Formula provides a visual depiction of previous fire-flow conditions in relation to present fire-flow conditions. These flow curves are very useful in identifying problem conditions and future needs. A comparative analysis of these water supply curves needs to be reviewed by the community water supply study group on a semiannual basis. This provides hard data on both the current condition of the water distribution system and specific locations where the water system needs to be improved or reinforced by optional water supply methods.

References:

1. Home and regional offices of the Insurance Services Office (ISO):

   - Home Office: New York City and Long Island
     ISO Building
     545 Washington Blvd.
     Jersey City, New Jersey 07310 1686
     Phone: 210-469-2000

   - Midwestern Region: Arkansas. Illinois, Indiana, Iowa, Kansas, Michigan, Minnesota, Missouri, Nebraska, North Dakota, Ohio, Oklahoma, South Dakota, and Wisconsin
     Lisle, Illinois
     Phone: 630-955-1080

   - Northeastern Region: Connecticut, Maine, Massachusetts, New Hampshire, Rhode Island, and Vermont
     Quincy, Massachusetts
     Phone: 617-770-3555

- Mid-Atlantic Region: Delaware, District of Columbia, Maryland, New Jersey, New York (upstate), and Pennsylvania

  Trenton, New Jersey

  Phone: 201-469-1000

- Southeastern Region: Alabama, Florida, Georgia, Kentucky, Louisiana, Mississippi, North Carolina, Puerto Rico, South Carolina, Tennessee, Virginia, and West Virginia

  Atlanta, Georgia

  Phone: 770-923-9898

- Western Region: Alaska, Arizona, California, Colorado, Hawaii, Idaho, Montana, Nevada, New Mexico, Oregon, Utah, Washington, and Wyoming

  San Francisco, California

  Phone: 415-439-4660

- State of Texas

2. AWWA Research Foundation/KIWA. *Impacts of Fire Flow Distribution System Water Quality, Design and Operation.* Sponsored by: American Water Works Association, 6666 West Quincy Ave. Denver Co. 80235-3098.

3. ibid. pg. 73.

4. Federal Emergency Management Agency. *Workbook–Alternative Water Supply: Planning and Implementing Process.*